The Comet Sweeper

The Comet Sweeper

*Caroline Herschel's
Astronomical Ambition*

Claire Brock

This edition published in the UK in 2017 by
Icon Books Ltd, Omnibus Business Centre,
39–41 North Road, London N7 9DP
email: info@iconbooks.com
www.iconbooks.com

Originally published in 2007 by Icon Books Ltd

Sold in the UK, Europe and Asia by
Faber & Faber Ltd, Bloomsbury House,
74–77 Great Russell Street,
London WC1B 3DA or their agents

Distributed in the UK, Europe and Asia by
Grantham Book Services, Trent Road,
Grantham NG31 7XQ

Distributed in the USA by
Publishers Group West,
1700 Fourth Street, Berkeley, CA 94710

Distributed in Canada by
Publishers Group Canada,
76 Stafford Street, Unit 300,
Toronto, Ontario M6J 2S1

Distributed in Australia and New Zealand by
Allen & Unwin Pty Ltd, PO Box 8500,
83 Alexander Street,
Crows Nest, NSW 2065

Distributed in South Africa by
Jonathan Ball, Office B4, The District,
41 Sir Lowry Road, Woodstock 7925

ISBN: 978-178578-166-7

Caroline Herschel, from an oil painting by Tielemann, 1829
(© National Maritime Museum, London).

The gold medal of the Astronomical Society of London (later the Royal Astronomical Society), awarded to Caroline Herschel in 1828. The telescope is William Herschel's 40-foot reflector, the symbol of the Astronomical Society; the motto of the Society, 'Whatever shines is to be noted down', appears above it. Isaac Newton is on the other side, with an excerpt from a Latin poem by Edmond Halley which appeared in the opening pages of Newton's Principia: 'the cloud [of ignorance] dispelled by science.'

For Ben Dew

Claire Brock is Associate Professor in the School of Arts at the University of Leicester. She is the author of *The Feminization of Fame* (Palgrave, 2006) and *British Women Surgeons and their Patients, 1860–1918* (Cambridge University Press, 2017), and the editor of *New Audiences for Science: Women, Children, and Labourers* (Pickering and Chatto, 2013). Claire Brock won the British Society for the History of Science's international Singer Prize (2005) and received a Wellcome Trust Research Leave Award (2012–14) for *British Women Surgeons and their Patients.*

Contents

Acknowledgements   x

INTRODUCTION   Astronomical ambition   1

CHAPTER I   Early life   13

CHAPTER II   Escape to Bath   61

CHAPTER III   From stage ornament to celebrated female astronomer   109

CHAPTER IV   Distinguished at last   163

Conclusion   219

Notes   225

Bibliography   265

Index   281

Acknowledgements

With grateful thanks to the following for all their generous support: Imogen Aitchison and Aaron Davies; Ann and the late Fred Brock; Siân and Paul Brock; Helen Brock and Joseph Giddings; Vera and Francis Connolly; Nicky Dawson and Simon Dew; Kathy and Chris Dew; Andy Lamb; and Julie Latham.

Simon Flynn of Icon Books and Jenny Uglow provided generous encouragement from the outset. Duncan Heath at Icon has been an astute reader and editor of the manuscript. Michael Hoskin's work on the Herschels has been inspiring; future scholars of the career of Caroline Herschel have him to thank for editing and making available Herschel's autobiographies.

Thanks are also due to the British Library for kind permission to quote from Caroline Herschel's correspondence. Herschel's intermingling of English and German, as well as her idiosyncratic spelling have been retained throughout. All translations from French or German texts, unless otherwise acknowledged, are my own.

Last, but certainly not least, I dedicate this book to Ben Dew, for everything and more.

Introduction

Astronomical ambition

At the beginning of August 1786, Caroline Herschel made the usual entries in her 'Book of Work Done'. With her brother William away, she was at leisure to survey the heavens, once she had completed her daily tasks. Very calmly, she entered the following:

> Aug 1. I have calculated 100 nebulae today, and this evening I saw an object which I believe will prove to morrow night to be a Comet.
>
> 2. To day I calculated 150 Nebulae. I fear it will not be clear to night, it has been raining throughout the whole day, but seems now to clear up a little 1, o'Clock the object of last night is a Comet.
>
> 3. I did not go to rest till I had wrote to Dr Blagden and Mr Aubert to announce the Comet.[1]

At the age of 36, Caroline Herschel had discovered her first comet. Just over a year later, in October 1787, with the award of £50 per year from George III, she would become the only woman in Britain to earn her living from the pursuit of science and, historically, the first woman to earn her living from astronomy.

Herschel's wages were ostensibly for assistance to her brother, William, whose discovery, in 1781, of the planet which would later become known as Uranus had propelled him on an unusual trajectory from a career in music at Bath to royal astronomer. Yet she was not simply an amanuensis or general dogsbody. Caroline Herschel made her own original findings, separate from the work she carried out for her illustrious brother. Her astronomical discoveries earned an international reputation and the highest accolades ever awarded, at that time, to a woman from the scientific community: a Gold Medal in 1828 and Honorary Membership in 1835 of the Astronomical Society of London. She was made an Honorary Member of the Royal Irish Academy three years later, and was presented with the Gold Medal of Science from the King of Prussia when she was in her nineties. Herschel's reputation was such that a letter written to her from the director of the Paris Observatory, Joseph Jérôme de Lalande, could be addressed simply to 'Mlle Caroline Herschel, Astronome Célèbre, Slough'.[2]

In 1844, at the age of 94, Caroline Herschel was

engaged in writing her memoirs for her nephew and his family. Looking back upon an extremely long life, she found herself exceptionally frustrated with her now useless body, the loss of her precious eyesight and, most vitally, her inability to be of use either to herself or to others. Typically, she phrased this by placing herself in brackets: 'I am so out of humour with myself at my inability at being of any farther use to any one; (or even to myself), that for these last three months I have not been able to add a single line to my Memoir.'[3] Despite physical feebleness, Caroline Herschel's mind was still acutely alert, and as she battled to force her body to keep pace with her brain, she assessed how her life could best be explained to her family. Although first offering the suggestion as a joke, Herschel repeated an intriguing idea more seriously in a second letter. What had started off as the mock fictional *Life and Adventures of Miss Caroline Herschel Solely for the Amusement of Lady Herschel* in April 1844 had become, by September of the same year, something far more fascinating. Fearing that she may not live to write anything substantial, Herschel directed her niece's attention to another possibility: '[This] may serve my grand Niece Arabella (perhaps;

with the assistance of some notes I found among the papers which my Nephew will find in the uper draw of my secretair) to twist into a Novel entitulet The Life and Adventures of Miss C. H. &c &c'.[4]

Although unable to carry on writing herself, Caroline Herschel was more than keen to see the story of her life handed down to posterity and preserved publicly for the benefit of future generations. A fictional treatment would secure the subject from instant identification, while simultaneously making only too clear whose life was being discussed. Caroline Herschel distrusted journalists and newspapers; as she so characteristically put it herself ten years earlier: '[I have been] looking over my store of astronomical and other memorandums of upwards of 50 years collecting and destreuing all what might produce noncence when coming through the hands of a Blockkopff in the Zeitungen.'[5] This way, her reputation could be managed posthumously by concerned and trustworthy family members, and compiled from her own papers.

Arabella Herschel never wrote the novel suggested by her great aunt. But Caroline Herschel had been right about one thing: her life, from its unpromising beginnings to its later, brilliant successes, made a perfect fiction. These fictional qualities have, however, also ensured that the subject has been suppressed by the legend. Until now, she has been treated almost exclusively as a dutiful sister to her more important brother. Much has been made about her selfless devotion to his studies, her long nights of waiting for his commands to write down stellar positions, her placing bits of food into his mouth when, due to excessive concentration on work, he had forgotten or was too tired to eat himself. This ceaseless support of William Herschel does indeed shine through in her memoirs, letters and diaries, which reveal sacrifice, stoicism, tireless labour and incredible self-abnegation. From the second half of the nineteenth century, Caroline Herschel's story has been told again and again, always with the same conclusions. For more than a century and a half, one viewpoint has prevailed. Comparing a late nineteenth-century assessment with a recent twenty-first-century analysis of Herschel's career, the similarities are striking. In 1895 Agnes M. Clerke's *The Herschels and Modern Astronomy* concluded the chapter on Caroline Herschel with the following summary:

[H]er faculties were of no common order, and they were rendered serviceable by moral strength and absolute devotedness. Her persistence was indomitable, her zeal was tempered by good sense; her endurance, courage, docility, and self-forgetfulness went to the limits of what is possible to human nature. With her readiness of hand and eye, her precision, her rapidity, her prompt obedience to a word or glance, she realised the ideal of what an assistant should be.

Herself and her performances she held in small esteem. Compliments and honours had no inflating effect upon her. Indeed, she deprecated them, lest they should tend to diminish her brother's glory.[6]

In 2004, Patricia Fara's examination of the place of women, science and power in the Enlightenment, *Pandora's Breeches*, traced Herschel's use of a canine metaphor to come to much the same conclusion as Clerke:

Male astronomers, [Herschel] wrote, were the huntsmen of science, while she was merely a pointer, eagerly awaiting friendly strokes and pats from her masters. [...] Like animals, men claimed, women were governed by their passions and needed to be controlled. In pictures they were shown together, twin models of fidelity and obedience to their master. Mary Wollstonecraft railed against the subservience exhibited by women like Caroline Herschel.[7]

And yet by invoking the eighteenth-century feminist writer Mary Wollstonecraft here, Fara points unconsciously to an area unexplored by historians of Herschel's career.

In her infamous polemic of 1792, *A Vindication of the Rights of Woman*, Wollstonecraft did indeed despise the fawning female, but her main concern was to draw attention to the reasons why women behaved in this way. For Wollstonecraft, the fault lay in education – or the lack of it – which rendered women trivial creatures, obsessed with physical appearance and eager solely to gain power through their sexuality. The only ambitions such ill-educated women aspired to achieve involved raising 'emotion instead of inspiring respect; and this ignoble desire, like the servility in absolute monarchies, destroys all strength of character'.[8] True ambition, claimed Wollstonecraft, manifested itself in the pursuit of reasonable pleasure and the conspicuousness of dignified virtue.[9] And at the basis of every virtue, Wollstonecraft placed 'independence'. Independence itself was both intellectual and financial; women would be able to succeed only in specifically female domestic tasks unless they became enlightened citizens, by earning their own subsistence and becoming independent from men in the same fashion as one man is independent of another.[10] For Wollstonecraft, virtue, independence and ambition formed a noble triumvirate.

Far from potentially despising Caroline Herschel for her canine attributes in the face of male superiority, Mary Wollstonecraft supported the virtues of female independence, and this accords only too well with Herschel's profoundly ambitious nature. From a very early age, Herschel expressed a desire to succeed in whatever she could. If, to re-employ the canine metaphors, she 'did nothing for [her] Brother than what a well-trained puppy Dog would have done',[11] she contributed fundamentally to her own training, first in music and then in astronomy, through a lifelong belief in the value of independence. Caroline Herschel may have followed her brother's instructions, but she was determined to achieve more than anyone could ask of her. Indeed, several of her cometary discoveries were made when she was alone. In this sense, ambition and independence, while still maintaining a dignified, virtuous appearance, were as much Caroline Herschel's desired *mode de vie* as they were Mary Wollstonecraft's. Perhaps the patience and persistence involved in 'minding the heavens' suited the contemporary character of the ideal woman, but Caroline Herschel did not carry out her observations in order to prove her stoic femaleness. As she informed the Astronomer Royal Nevil Maskelyne in September 1798: 'I do own myself to be vain, because I would not wish to be singular; and was there ever a woman without vanity? or a man either? only with this difference, that among gentlemen the commodity is generally styled ambition.'[12] Unlike other women, Caroline Herschel's discoveries had allowed her to direct her ambitions to worthy causes.

Rehabilitating Caroline Herschel will involve not only reconsidering women's place in the history of science through an investigation into the career of a woman who actually made scientific discoveries, but also examining how eighteenth and early nineteenth-century women could exercise their ambitions in a society which forbade them political representation. While historians must beware of exaggerating, distorting or overstating the lives of scientific women and converting them into feminist heroines,[13] this is certainly not the intention of this book. The career and writings of Caroline Herschel offer an insight into the position of women at the time, as well as revealing that women were not only able to understand the 'harder' physical sciences of mathematics and astronomy, but also to participate in their progression through original discovery and explication, both to specialists and to larger public audiences. In order to evaluate Caroline Herschel's place in the history of science, therefore, it is necessary to examine her position within eighteenth- and early nineteenth-century society. All important are the reactions of her contemporaries and, most fundamentally, how Herschel herself conceived of her own social, cultural and scientific role. By filling in the gaps and allowing Herschel to speak for herself, it will be possible to gauge how this one woman had the opportunity to make her indelible mark upon the nascent scientific community of her time.

Work, independence and ambition mattered enormously to Caroline Herschel, but so did astronomy. Her embarrassment at the awards she received nearly 30 years after her last original discovery was not shame at the publicity, but rather irritation that her age had prevented her from living up to her distinction in recent years. Alert still in her eighties and nineties, Caroline Herschel desired to remain involved in the astronomical world, whose news kept her living with 'morsels [...] to feed upon'.[14] Nor was she averse to advising her nephew where in the heavens to investigate stellar formations, annoyed that she could not join him in his sweeps. Her death in January 1848 at the age of 97 brought to a close one of the most extraordinarily varied and successful careers of the late eighteenth century. In Caroline Herschel's own estimation, she had never achieved enough, wasted too much time, never become as independent as she would have wished. But for others, including the members of the Astronomical Society of London, she had more than fulfilled her astronomical ambitions, both as an assistant and, most importantly, as a recognised astronomer in her own right. Reading through his aunt's autobiographical account in May 1827, John Herschel informed her that she 'under-rate[d] both the value and the merit of [her] own services in [William Herschel's] cause', but that this was counterbalanced by her deserved reputation. The world, indeed, did Caroline Herschel more justice.[15]

Chapter I

Early Life

When she looked back upon her life, Caroline Herschel found the period of her childhood brought back the most painful memories. In a letter written to her nephew John Herschel's wife Margaret in September 1838, she offered a reason why her youth had been joyless and unpleasant: 'But as it was my Lot to be the Ashenbröthle of the Family (being the only Girl) I never find time for improving myself in many things'.[1] At the age of 88, the frustrating fact that Caroline Herschel had been prevented from following her ambitions of improvement still rankled with her profoundly. Forced into becoming what can best be translated as a Cinderella figure, she was compelled to perform mundane household tasks while others went to the ball. For Herschel, the ballroom was something far more important than an evening soirée. It was something more abstract – a room without walls, indeed. To go to the ball would mean a chance to escape from household drudgery and obtain independence.

Carolina Lucretia Herschel[2] was born on 16 March 1750 in the German city of Hanover, where she would live for the first 22 and the last 26 years of her life. The eighth child of ten, six of whom survived to adulthood, Caroline Herschel was very low on the list of parental priorities. Her only surviving sister, Sophia, was almost seventeen years older, and married when Caroline was only five years old. Thus, Herschel's suggestion that her slave-like existence was founded on her femaleness is ostensibly true. With four older brothers – the eldest of whom, the detested Jacob, was almost sixteen when his youngest sister was born – and one younger, it is only too easy to see how Caroline was compelled to learn the womanly duties expected of her before her time. Her intense dislike of Jacob was mirrored by the irritation she felt for her illiterate and overbearing mother Anna. A ray of hope in Caroline Herschel's young life was her father Isaac, whom she worshipped, and who never ceased to encourage an intelligent daughter always eager to profit from his lessons. The affection for her beloved brother William, to whom she would later devote her time and energy by acting both as a concert singer and an assistant in astronomy, is also apparent from the outset. Caroline loved only those who encouraged and distinguished her, for, by the rest of the family, and especially by Jacob and by Anna, her desires were either ignored, suppressed or mocked.

The socially and intellectually unequal marriage of Isaac and Anna Herschel contributed undoubtedly to their youngest daughter's divided loyalties. Isaac came from a creative if autodidactic family, and revelled in the pursuit of music and art, as well as mathematics. His background was quite lowly: his own father Abraham was a gardener in the fashionable ornamental pleasure gardens in Hohentziatz, while his mother was the daughter of a tanner from Loburg. Their pretensions to higher status were signalled by the classical names given to their two elder children, Eusebius and Appolonia. Isaac, the youngest, was expected to follow in the family trade as a gardener, but a love of music distracted him from his future career; not the last time a Herschel would wilfully 'try' something else. Caroline and William Herschel were certainly their father's children in this respect.

Losing interest in horticulture, Isaac taught himself to play the oboe and soon perfected the instrument enough to find permanent employment in the musical profession. Finally settling in Hanover after a peripatetic existence trying to find a post in Prussia, he joined the band of the Hanoverian Foot Guards in August 1731. At the age of 24, Isaac was ready to start a new life with a job in which he displayed talent and, most importantly, one which he enjoyed.

A year later, he met and married Anna Ilse Moritzen, who was born and brought up only three miles from Hanover. Anna's provincial outlook was matched by her lack of education, and such an unlikely marriage with the aspiring Isaac can be explained by the fact that Anna was already pregnant with their first child Sophia when they married in October 1732.[3] Caroline Herschel and her siblings were thus brought up by two very different parents: one, artistic and hard-working; the other, illiterate and lazy. Isaac's sense of purpose, his generous desire to aid his children in their interests, can be contrasted dramatically in Caroline Herschel's writings with Anna's meanness. While Isaac was constantly doing something and filling his time with useful tasks, Anna never actually does anything, except visit friends[4] and stunt Caroline's emotional and intellectual growth. Although Isaac had not been educated to any but the most moderate standard, or even that far beyond his wife, his belief in his children's abilities was suffused with the pride and ambition of the autodidact. In her memoirs Caroline Herschel remembered primarily the expansiveness of her idolised father, and compared it continually and scathingly with the narrow-mindedness of her mother.

Anna sought to hold on to her unpaid housekeeper by preventing her youngest daughter from following any interests outside the home as a child. Caroline seems to have been a burden for her mother, who frequently sent her on useless errands and then berated or simply forgot about her. One searing recollection occurs in both of the autobiographies Herschel wrote for her family later in life, and particularly upsets her. Fearful of conflict with the French, the King of Great Britain and ruler of Hanover, George II, ordered the Hanoverian Guards to Britain as precautionary reinforcement troops in March 1756; a step that proved fortuitous, as what became known as the Seven Years War would begin between the two countries in May.[5] Isaac Herschel and his two elder sons, Jacob and William, who had followed their father into the military band after leaving school, were sent with the Guards.[6] As he had requested a discharge from the band in order to attempt to further his musical career in more congenial surroundings, Jacob returned to Hanover in the autumn of 1756; his father and younger brother would follow a few months later. Aware of their impending return, Anna, who was preparing a special homecoming meal, sent the six-year-old Caroline out to search for Isaac and William. The determined little girl searched endlessly, despite extreme cold and tiredness; her desire to fulfil all her tasks apparent even at this very early age. In her second autobiography, written in her eighties, the elderly woman remembered painfully the double discomfort of a stomach ache to add to her considerable woes.[7] Returning home disconsolate, Herschel was astonished to witness the whole family merrily tucking into their dinner. In both accounts, only her beloved William welcomed her home. The others carried on with their meal, 'nobody greeting' the frozen child, as they were 'too happy in the Thumult of their joy at seeing one another again'. Herschel remembered mournfully that her 'absence had never been perceived'.[8]

The little girl seemed a constant irritation to her mother. When she was not at school, she was forced out of the house by Anna and by her older sister Sophia, whom Caroline had never met before she moved in with her parents when her husband, Griesbach, also a member of the band of the Hanoverian Guards, accompanied his father and brothers-in-law to Britain. Sophia despised 'having children about her', and her little sister was easily dispensed with. In her old age, Herschel recalled being sent to 'play *auf dem Walle*' with her brother Alexander, who was five years her senior, or with the neighbours' children. As she was so young, Herschel was unable to join in with their games, and was compelled to hang around on the embankment and watch her brother skating 'till he chose to get home'. Isolated and friendless, the little girl simply stood and shivered. 'In short', she claimed, with some understandably succinct bitterness, 'there was no one who cared anything about me'. Tellingly, 'anything' of the finalised version had replaced 'much' of the original draft.[9]

With her father and favourite brother absent, Herschel noted how she was compelled to act as a footman or waiter for her mother, who would frequently entertain their landlord. At around seven years old, she was, of course, incapable of balancing plates or waiting efficiently on adults. And yet this inevitable childish deficiency did not stop Anna giving her daughter 'many a wipping for being too awkward',[10] or, in fact, using her to finish off her own work when she was unable or unwilling to complete it herself. With Sophia recovering from the birth of her first child and Anna 'comforting' her eldest daughter, the eight-year-old Caroline was left to complete the task of 'furnishing the Army with Tents and Linnen'. After a day at school, the little girl returned to wear her fingers to the bone sewing for her mother. Herschel did not mean this metaphorically. Her small hands were simply unable to use a thimble. Despite the physical suffering caused by all the needlework, Caroline was determined to succeed at her task, and stuck to it doggedly, touchingly performing it 'with all [her] might' to prove herself able and useful.[11]

In the parts of *The First Autobiography* originally written for her younger brother Dietrich, Herschel's upset at this loveless first decade of her life caused her literally to break down mid-sentence. Describing Jacob's attempts to hide from military press-gangers when Hanover was occupied by the French during the Seven Years War, Herschel started to recount the following: 'The next time I saw him was when he came running to my Mother with a letter, the contents of which …' Breaking off abruptly, Herschel notified Dietrich that: 'After reading over many pages, I thought it best to destroy them, and merely to write down what I remember to have passed in our family at home and abroad.'[12] The only way Herschel could continue to narrate her life, she suggested here, was to stick rigidly to the facts; the feelings were as searing as the flames which destroyed the memories.

If her domestic surroundings caused Herschel distress, her geographical location frustrated her equally. When she returned to Hanover after William's death in 1822, Herschel had nothing but contempt for the Hanoverian people. In letters written in the 1830s to her nephew and his wife and other correspondents, Herschel expressed disdain for 'this dissipated City', 'this Horrible Hannover'.[13] And it was not only the city that the elderly Herschel despised. The people were pleasure-seeking and vain: 'neither Man, Woman or Chield in Han. to be found but they must spend the evening at Balls, Plays Routs, Clubs &c and not a Month goes over our Heads, without a Jubily is celebrated at enormous expenses'.[14] '[S]tupid Hannoverians' caused '*Botheration* and intrusion'; they were a 'mongrel breed'.[15] In a defiant exclamation, Caroline cried: 'I will be no Hannoverian!'[16] Despite the fact that Hanover was her birth-place and the city in which she had grown up, Herschel felt no ties to the place that she linked inextricably with servitude and oppression. Portraying her own country in this way, Herschel sought to distance herself from contemporary assumptions about Germanic people as dull, dreary and domestic. The British, French and Italians especially looked down upon the Germans, and statistically the German states appeared backward, with a 75 per cent peasant population in the eighteenth century.[17] This figure compared unfavourably with around 60 per cent in France and 30 per cent in Britain.[18] The German novelist and author of the sentimental fiction *The History of Lady Sophia Sternheim* (1771), Sophie von La Roche, indirectly expressed how many other

Europeans viewed Germany when she visited Britain in 1786 and commented upon the progress of print culture in comparison to that of her own nation:

> At home we think we have done a great deal for the common man by inserting a modicrum of good sense in the calendars, which are only issued to the people annually; but in England and in London there are 21 daily newspapers, containing news of foreign parts and states and excellent articles on all kinds of subjects, poetry, humorous and witty passages, satires and moral maxims, historical and political essays in addition. I already mentioned the Ipswich paper at Mistress Norman's in Helveetsluys on that account, for this is only a provincial town, and yet so many ideas for one's enlightenment are contained in it.[19]

In the periodical the *Deutsche Chronik*, Christian Daniel Schubart informed his readers exactly what the rest of Europe thought of them. In 1774, he noted the proud superiority of the British, who viewed the Germans as subhuman and feared their own monarch travelling in such barbarous lands (this is, ironically, in spite of the fact that from 1714 to 1837, Great

Britain was ruled by Hanoverians). German travellers were disbelieved, and astounded their British hosts when they told them that freedom existed in what they believed was a terrifying and hellish country. The ignorance about German customs and culture was not simply confined to the uneducated or the illiterate. Even Jonathan Swift, author of one of the bestselling works of the eighteenth century, *Gulliver's Travels* (1726), found the Germans to be the 'most stupid people on earth'.[20]

It is not even historically accurate to discuss eighteenth-century Germany or a particularly German character in this period, because as a geographical entity, Germany simply did not exist. If Caroline Herschel and others did not feel pride in their nation, it was because that nation was made up of 294 states or 2,303 territories and jurisdictions which comprehend those areas known now as Germany.[21] Hanover suffered a double dislocation, as, from 1714, its ruler was also King of Great Britain. Thus, for most of the eighteenth century, Hanover was the only German state without a resident prince. While George I visited Hanover an amazing six times in a reign of thirteen years, and George II twelve times in 27 years,

George III expressed similar views to Caroline Herschel about a state that he viewed as a parasitic burden: it was a 'horrid electorate, which has always lived upon the very vitals of this poor country' and proved 'an enormous expense'.[22] After the British and Hanoverian defeat at the Battle of Hastenback in August 1757, Hanover was occupied by French troops until February 1758. Herschel described this traumatic time as 'depressing': 'we were almost immediately in the power of the French troops, each house being crammed with men. In that where we were obliged to bewail in silence our cruel fate.'[23] The occupation indeed signalled the declining British interest in Hanoverian fortunes, and it took the army of Frederick the Great of Prussia to defeat the French and restore Hanover. George III's waning desire to continue to bolster Hanover was reflected in the Peace of Paris and the subsequent Treaty of Hubertusburg which ended the Seven Years War in 1763, whereby both Britain and France effectively turned their backs on potential and actual German possessions. This was due, largely, to financial concerns. The British wanted an alliance which would protect Hanover from future invasion, but the central and eastern European powers would not defend the electorate for nothing. From now on, Hanover was on its own.[24]

Historians disagree as to whether this absence of a central powerful figure caused economic and cultural stagnation or in fact encouraged a more laissez-faire attitude to commerce and culture.[25] All agree, however, that Hanover's apparently headless state retarded further progress. The electorate had potential: a coastline on the North Sea, and access to the major rivers, the Weser and the Elbe, which reached into the heart of the German states. But these possibilities went unexplored. The coastline remained ill-adapted to economic development and the mouths of the rivers were not Hanoverian, but were dominated by Bremen and Hamburg – Imperial cities. External trade was weak and Britain did not encourage preferential trading treatment. Economically speaking, Hanover, with its declining exports of raw linen, had little to offer the increasingly powerful British industries.[26] The opportunities which may have existed for economic and political development in eighteenth-century Hanover required the leadership and attention of a strong executive authority; but the departure of George I for Britain in 1714 effectively destroyed the possibility of progress and improvement within the state.[27] Caroline Herschel saw herself as doubly disadvantaged. Prevented from improving herself at home, she was also confined to surroundings which were similarly noted for their lack of progression.

Formal education might have offered a potential outlet for Herschel. Paradoxically, and in spite of the attitudes of other Europeans, German education was among the most advanced and progressive in Europe during the eighteenth century. Statistically, literacy levels were also impressively high. By 1800, about 65 per cent of French men and 35 per cent of women could sign their names, and probably more could read. Britain, Holland, Scandinavia and parts of Germany (Protestant areas in the North) registered a higher rate. It is possible, indeed, that the highest literacy levels in Europe may have been reached in Prussia, because of the introduction of obligatory elementary schooling from 1717.[28] The British may have scoffed at the ill-educated, boorish Germans, but schooling in eighteenth- and early nineteenth-century Britain, despite high literacy rates, was notoriously poor, especially for girls. In *A Vindication of the Rights of Woman*, Mary Wollstonecraft condemned the regimented approach to teaching children in single-sex schools in Britain, whereby they were forced to 'recite what they did not understand', in 'parrot-like prattle'. Women were educated solely for show, 'obliged to pace with steady deportment stupidly backwards and forwards, holding up their heads and turning out their toes'.[29] As a schoolchild in Musselburgh, Scotland in the 1790s, boarding at the benevolent-sounding Miss Primrose's School, Mary Somerville noted the focus on the trivial and showy:

> My future companions, who were all older than I, came round me like a swarm of bees, and asked if my father had a title, what was the name of our estate, if we kept a carriage, and other such questions, which made me first feel the difference of station. ... A few days after my arrival, although perfectly straight and well-made, I was enclosed in stiff stays with a steel busk in front, while, above my frock, bands drew my shoulders back till the shoulder-blades met. Then a steel rod, with a semi-circle which went under the chin, was clasped to

the steel busk in my stays. In this constrained state I, and most of the younger girls, had to prepare our lessons. The chief thing I had to do was to learn by heart a page of Johnson's dictionary, not only to spell the words, give their parts of speech and meaning, but as an exercise of memory to remember their order of succession. Besides I had to learn the first principles of writing, and the rudiments of French and English grammar. The method of teaching was extremely tedious and inefficient.

Unsurprisingly, the intelligent Somerville remained at the school for a year before fleeing and feeling as if she was a 'wild animal escaped out of a cage'.[30] Fictional representations of the poverty of female education are only too apparent in novels of the late eighteenth and early nineteenth century. The barely literate Harriet Smith of Jane Austen's novel *Emma* (1816) attends a school where 'girls might be sent to be out of the way and scramble themselves into a little education, without any danger of coming back prodigies', while Charlotte Jennings of *Sense and Sensibility* (1811), whose only 'proof of her having spent seven years at a great school in town to some effect' was a 'landscape in coloured silks of her performance' which 'hung over the mantelpiece in her old room'.[31]

By contrast, and unlike most other European countries, German school systems came into their own when Caroline Herschel was at school and after, in the 1760s and 1770s. At an elementary level, German schools in Protestant areas were therefore becoming more successful at teaching the basics to more and more children. Under the new system, parents were provided with help in paying school fees, with the poorest allowed to send their children to school at a lower rate or even free of charge.[32] From the ages of five or six to thirteen or fourteen, children were legally obliged to attend school. But now that education had become compulsory, what did this actually mean in practice for eighteenth-century German children? The institutional framework may have been in place, but did Caroline Herschel or other girls of her age and social status benefit from the changes, or was formal education for females as chronically impoverished as that of Mary Somerville or the fictional Charlotte Jennings?

For a start, girls were still taught differently from boys, and this usually meant that they were taught the basics soundly but were not required or encouraged to improve themselves beyond being able to read or write. This was, however, improving in some parts of Germany. Beyond the compulsory requirements of reading, writing and religion, some children received music lessons or were taught how to sing. In the more advanced cities and states, girls were even taught arithmetic, while Eisenach specifically excluded girls from learning even the most basic mathematics. We do not know for certain if Herschel learnt arithmetic at school, in line with the developments in this field for girls, but it seems possible that she was taught the bare minimum, as William was forced to re-educate his sister mathematically in later years.[33] Hanover did, after all, offer a pioneering middle-school education for girls at the Hof- und Töchter-schule, or institution for daughters – the future housewives and mothers – which expanded intellectually and practically an elementary education. The school continued teaching the basics, alongside the making of handicrafts, but with the addition of French, history, geography and the natural sciences, as well as arithmetic.[34] So it is possible that this widened curriculum may have trickled down to the elementary schools in the electorate. The more advanced states of Halle, Prussia and Saxony included the rather vague-sounding 'other arts and sciences' at elementary level.[35] The example of Eisenach reveals that not all states showed enthusiasm for educational reform, and improvements in girls' schooling was a particularly contentious issue.

Three dominant notions about educating women circulated around eighteenth-century Germany, as they did around most of Europe at the time.[36] Firstly, as women were destined solely to be wives and mothers and thus be responsible for running a household, they needed only be educated in order to carry out such tasks efficiently. This idea could be channelled into another notion prevalent at the time: that men and women had their own special abilities in different areas and should be educated accordingly. Both theories could work one of two ways for women. Either it meant that their education could be expanded to include the necessary arts and sciences needed to maintain an efficient domesticity, or it could condemn women solely to practical tasks or social skills:

there was certainly no need for intellect with such a destination. Several female educational reformers had employed the former tactic: it was woman's duty to be an efficient housekeeper. Some knowledge of the various forms of natural philosophy was deemed intrinsic to women's development. From Mary Astell and Bathsua Makin in the late seventeenth century to Maria Edgeworth and Mary Wollstonecraft in the 1790s, scientific *savoir faire* was considered compatible with domestic duties and the smooth running of an efficient household. Awareness of the rudiments of mathematics could encourage economy, and knowledge of the laws of chemistry would improve the art of culinary accuracy. As Mary Astell noted in *The Christian Religion* (1705):

> And since it is allow'd on all hands, that the Mens Business is without Doors, and theirs is an Active Life; Women who ought to be Retir'd, are for this reason design'd by Providence for Speculation: Providence, which allots every one an Employment, and never intended that any one shou'd give themselves up to Idleness and Unprofitable Amusements. And I make no question but great

Improvements might be made in the Sciences, were not Women enviously excluded from this their proper business.[37]

Nearly 90 years later, Mary Wollstonecraft agreed. Rather than distract women from their particularly female domestic duties, pursuit of the sciences could only improve women's dedication; a steady eye could only strengthen the mind.[38]

But pursuit of scientific knowledge need not ensure that one became confined within the home. Botany and entomology proved especially enticing for women of this period.[39] Not only was the study of nature considered acceptable for female education, it was actually encouraged as a way of learning about the surrounding world which was cheap (readily available material) and healthy (expeditions required physical exercise). As a male contributor to Eliza Haywood's periodical *The Female Spectator* noted, women could excel in these pursuits, each with the potential to become a new botanical Columbus. The possibility, as the contributor put it:

To have their names set down on this Occasion, in the Memoirs and Transactions of [the Royal

Society], would be gratifying a laudable Ambition, and a far greater Addition to their Charms than the Reputation of having been the first in the Mode, or even of being the Inventress of the most becoming and best fancied Trimming or Embroidery, that ever engrossed the Attention of her own Sex, or the Admiration of ours.[40]

For some, intellectual studies were wholly inappropriate for the female mind in the first place. Natural philosophy, especially, should be pursued only so far and viewed solely as a leisurely, fashionable activity. Women certainly should not apply themselves as diligently to telescopes and microscopes as they did to more appropriate female occupations, such as household tasks. In *Recherche de la vérité* (1674), the French philosopher Nicolas Malebranche stated forcefully that women's minds simply were not formed to cope with the abstract nature of mathematics or the physical sciences: 'Everything abstract is incomprehensible to them. They cannot use their imagination to develop compound or awkward questions. They only consider the surface of things, and their imagination is not strong or extensive enough to plumb the depths.'[41] In one of the most important educational treatises of the eighteenth century, *Emile*, Jean-Jacques Rousseau commented specifically on the disparity between a female character and the harder sciences. Women do not possess 'sufficient precision and attention to succeed at the exact sciences. And as for the physical sciences, they are for the sex which is most active, gets around more, and sees more objects, the sex which has more strength and uses it more to judge the relations of sensible beings and the laws of nature.'[42] And even the supportive contributor to the British periodical *The Female Spectator* warned his audience about the 'severe and abstruse' complexities of the physical sciences. Persistence was required: 'a Depth of Learning, a Strength of Judgment, and a Length of Time, to be ranged and digested so as to render them either pleasing or beneficial.'[43] But, as Rousseau and the male periodical contributor realised, the eighteenth-century female craze for the abstract had the longevity of seasonal fashions. After all, while women could attempt equations, they had insufficient rigour, precision and intellect to devote themselves to mathematical studies. To acquire scientific skill beyond superficial and sociable appreciation was to deprive men of enlivening company and women of the understanding of the rules of civility. Scientific ideas were to be cultivated and understood as a necessary section of contemporary culture, but only as a small part of a female education.

Finally, some argued that women were definitely capable of succeeding in a more complex educational atmosphere. Even the philosopher Nicolas Malebranche had claimed that there were some exceptions to the rule that female brains were physically incapable of coping with serious matter.[44] But some then maintained that women should not be educated in such a manner – what was the ultimate point in teaching women arts and sciences that they could not put to use in everyday life? The continuing arguments over the correct mode of female education both helped and hindered the cause of women's intellectual improvement.

Like most of her contemporaries, including her brothers, Caroline Herschel went to school until the age of fourteen. In her writings she does not give too much information about her schooldays, but it is easy to picture what they were like from what she does not say, and, indeed, how much she had yet to learn once she left. Herschel claimed that she 'went dayly with [her] little brother to School by the Garrison Küster' and returned 'from thence at three in the afternoon'. She then went alone for another three hours to a school where she 'learned to knit'. Extra schooling in activities such as knitting would have cost Herschel's parents more money, but it is clear that her mother at least felt this was a well-invested opportunity for further exploiting her daughter around the home. Anna was willing to contribute to her daughter's educational opportunities only when she could profit from them. Indeed: 'from that time on I was fully employed with providing my Brothers with stockings.' In a typical moment of self-deprecating humour, which also reveals how hard Herschel desired to work in order to improve herself, she notes: 'the first pair for Alexander touched the floor when I stood upright finishing the foot.' The contrast between Anna's generation and her daughter's is highlighted when Herschel describes how highly prized (and exploited) her literacy was to the surrounding neighbourhood. At the age of seven or eight, Caroline Herschel was penning 'not only [her]

Mothers letters to [her] Father but to many a poor soldiers Wife in our Nighbourhood to her Husband in the Camp'.[45] This fact, in itself, reveals how compulsory elementary education had aided Caroline Herschel and her peers to aim higher than their mothers ever could.

The only time in which school is discussed at any length in the autobiographies is when Herschel was effectively preparing to leave. While both boys and girls left elementary education around the age of fourteen, it was considered a particular rite of passage for the latter. Many of the major German states, clearly including Hanover from the Herschel children's experience, linked graduation from elementary school with permission for both sexes to participate in religious confession or confirmation. For girls, however, the significance of leaving school took on another specifically gendered dimension. In eighteenth-century Germany, the fact that girls left school at thirteen or fourteen signalled that they were now sexually and intellectually mature and thus able to marry.[46] And even though the average age for marriage was at least ten years above the school leaving age, as was the case in most of Europe in the eighteenth century,[47] this link was maintained in theory if not in practice. From the age of fourteen, girls were encouraged to think about their future roles: as wives and mothers. They could improve their minds no further, but they could learn the practical skills to enable them to perfect their allotted role in life.

Caroline Herschel's coming of age was certainly not the momentous rite of passage that other girls experienced. For a start, she had not only to attend school and church, but was compelled to perform the endless household tasks that Anna placed before her. She referred to being continually 'employ'd in the drudgery of the Scullery', and compounded the sense of her servitude by noting that she was frequently unable to 'make one in the groope when the Famely were assembled together'.[48] Like the lowliest scullery maid, Caroline was often divorced physically and therefore socially from her own family. When the time for her confirmation came, the extremely touching pride in her exciting and novel situation was countered by the profound indifference of the rest of the family. Excepting William, no one else bothered to congratulate their daughter or sister on her special day: 'Sunday the 8th was the (to me) eventful day of my Confirmation and I left home not a little proud and encouraged by my dear Brother W[illiam]'s approbation of my appearance in my new Gown &c.'[49] The brackets surrounding her own feelings and her apparent experiencing of the event on her own reveal only too easily how unimportant she was to the majority of her family.

An explanation for the family's lack of interest in Herschel's maturity can be found in the attitude of Isaac and Anna to their youngest daughter's dire future prospects. The doubly important events of confirmation and leaving school, as already noted, took on an additional significance for girls. For most parents, the sexual maturity of their daughters enabled them to begin the expected route to marriage. From a very early age, however, Caroline Herschel had been informed that she would have little chance of ever finding a husband. An intimation of her unsuitability came very early from an unexpected angle: her usually more subtle father. Isaac's warnings obviously had a devastating impact upon the young girl:

I never forgot the caution my dear Father gave

me; against all thoughts of marr[y]ing, sa[y]ing
as I was neither handsom nor rich it was not likely
that any one would make me an offer, till perhaps
when far advanced in life some old man might
take me for my good qualities.

Scrupulously practical, Isaac ensured that his youngest daughter could never delude herself as to the
'glumy prospects of the future'.[50]

Herschel's looks had been affected at the age of
four by a smallpox attack which killed her brother,
Frantz Johann, before he was two years old. Herschel
was lucky. Smallpox was an appalling disease which
killed thousands across Europe before inoculations
were introduced en masse at the end of the eighteenth century.[51] In the same year that Herschel con-
tracted the disease, the popular British periodical, the
Gentleman's Magazine, offered its monthly figures
concerning births and deaths in the metropolis: 'The
London General Bill of Christenings and Burials from
December 11, 1753, to December 10, 1754.' During
the year, 14,947 babies were christened, compared
with 22,696 burials. Of this figure, 6,115 had died
from unspecified 'convulsions' and 4,241 from con-

sumption. Over 10 per cent of recorded deaths were from smallpox, which claimed 2,359 lives; 413 more than had died from old age.[52] But while Herschel was one of the more fortunate survivors of the disease, it seems that she suffered the other indignity of the smallpox sufferer: permanent scarring. Life, death and suffering are linked together in her narrative of the time:

> The 16th of March 1750 I was born, and in 1752 another Son; who died 1754 of the small pox which I had at the same time; and though recovered, I did not escape being totally disfigured and suffering some injury in my left eye.[53]

Her physical appearance blighted from the age of four, Caroline Herschel grew up almost certain of her fate as a lonely spinster.

It was a combination of factors working against Herschel, however, which prevented her marriage. Physical deformities were more than common in an age when many suffered from the disfigurement caused by an all too prevalent disease. With her lowly domestic status as Anna's unpaid servant, when could Herschel find either the time, the inclination or the finances to join social gatherings with her much older brothers? The glittering possibilities of 'polite society' for this Cinderella figure are seen only in second-hand form – from Dietrich's attendance at musical parties: 'I was not only entertained but frequently instructed by his relation of what was going forward in learned and polite societies.'[54] And, as Isaac noted so brutally and yet so honestly, marriage was frequently an economic transaction and he had nothing to give her. As Mary Wollstonecraft informed her readers in *A Vindication of the Rights of Men* (1790), marriage in the eighteenth century was little more than 'legal prostitution to increase wealth or shun poverty'.[55] Perhaps Caroline Herschel was making a lucky escape from the inevitability of matrimony, for marriage was scarcely held sacred, contemporaries suggested, in the German states. Northern Germany was notorious for the ease of obtaining a divorce. Unlike the rest of Europe, where this process was difficult, if not impossible, except for the very rich, it seemed all too simple for the Germans. In her social and cultural investigation of Germany, *De l'Allemagne*, eventually published in London in 1813 after having been suppressed by Napoleon three years earlier,

Germaine de Staël stated that in Protestant areas: 'One can change one's spouse there as peaceably as if it were a matter of arranging the incidents of a play.'[56] Without beauty or wealth to recommend her, Caroline Herschel looked set to join the ranks of the despised spinster class before she had even reached maturity.

But, forewarned as she was, Herschel was set to make the most of her situation. If she could not marry, then she would improve herself in order to achieve independence. To be 'useful' was Caroline Herschel's greatest desire. And yet it was not simply to be of use to others that made her determined to better her prospects. There was a clear self-motivation to Herschel's theory of usefulness. Even into old age, she clung to this mode of self-sufficiency. To help others, ultimately, was to help oneself. This sense of duty, to herself and to others, was inculcated at a very early age by her domestic surroundings. Anna and Isaac's eldest son Jacob was a profligate and a spendthrift, who had the Herschel trait of self-improvement trained solely on gaining a higher social position. When she was eleven or twelve, Herschel noticed how 'incorigible' Jacob was 'where Luxury, ease and

Ostentation were in the case'. As well as overspend-
ing on clothing and parties in order to show off the
wealth he did not possess, Jacob, although a genuine-
ly talented musician, thought too highly of his abilities. Exasperated by his smug and arrogant nature,
his youngest sister cried:

> [W]hen he had a Quartet party where his new
> Overtures and other compositions were tried and
> admired, which my Father expected and hoped
> would be turned to some profit by publishing
> them; ... there was no printer who bid high
> enough, and nothing was published![57]

Isaac was evidently concerned for his children's
future, and this fear that they would not amount to
anything was passed straight to his sensitive daughter. With his health declining every day, lamented
Herschel, her father could see 'no prospect of any
one of his children capable or even willing to help
themselves'.[58]

But one of Isaac's children was more than willing
to help herself, no matter what the difficulties, and
this child was, of course, Caroline. This trait was much
in evidence during her school years, when she proved

herself to be the autodidact Isaac's daughter. German girls were encouraged to sit quietly (*stillsitzen*) while remaining conspicuously active; certainly an advantageous occupation for the future astronomer. From *stillsitzen* derives the German word *Sittsamkeit* – demureness or modesty.[59] In a strident, anonymous English text of 1758 entitled *Female Rights vindicated, or the Equality of the Sexes Morally and Physically proved*, the author noted that women were innately suited to scientific study, and placed physical sciences at the top of the list. In order to study astronomy, mathematics or physics one needed only to be closely observant, with an eye for detail, and be possessed of much patience and energy. Astronomy especially required only good eyesight:

> As to Astronomy, what is there the Women may not equally discover with the Men? All the planetary Phaenomena are visible to them by the Use of Optics; and a proper Attention to their periodical Revolutions, will inform them of their Motion and their Course. Perhaps, their general Ignorance of Mathematics may be urged as a Reason for their Conclusions being vague, and their

Problems incorrect; but this does not prove their Inability of acquiring a Knowledge therein, since all their Faculties are clearly evinced to be as perfect as those of the Men.[60]

By making several comparisons with the difficulty of embroidery and other female skills, the author suggested that the characteristics required for the successful astronomer could be found in any female educated to sit still and embroider or sew for hours on end.[61] *Female Rights vindicated* concluded confidently that: 'All the Sciences have been investigated to their greatest Depths by Females.'[62]

Caroline advanced herself intellectually while no one noticed. Several times in her autobiographies, she revealed how she literally listened and learnt. German educational writers suggested that girls should employ their leisure time with traditional handicrafts like spinning, sewing or weaving.[63] Indeed, even Herschel admitted that her narrow-minded mother was an excellent and diligent spinner.[64] Of course, the recommended tasks were hardly recreational pursuits. Not a minute was wasted; even 'hobbies' for German women were fundamentally profitable for the family. In order to make some more money, Isaac Herschel took in paying pupils, as well as educating his own children, who had all inherited his musical talent. As a child, Herschel recalled that she was

> frequently called to join the 2$^\text{d}$ Violin in an Overture, for my Father found pleasure in giving me sometimes a Lesson before the Instruments were laid by after practising with Dietrich, for I never was missing at those hours, sitting in a corner with my Knitting and listening all the while.[65]

Jacob's obsession with extravagantly entertaining guests who lived 'in the most luxurious stile', however, frequently interrupted the domestic harmony that existed between Isaac and his youngest daughter, as did Anna's disapproval – a fact that Herschel remembered into old age. In a letter written in September 1838 to Margaret Herschel, her nephew's wife, Caroline laughed at the subterfuge in an afterthought: 'NB when my mother was not at home, Amen!'[66] She bewailed these interruptions to her progress with scarcely disguised scorn:

[M]y Father was deprived of the pleasure he found in hearing me read to him whilst he was copying music, and my brother Dietrich lost many afternoons leson of which loss I participated in no small degree; for I alwa[y]s was seated with my knitting as near as I could get ... But our appartments were too close together for allowing any practising when there was company with my brother and my reading was also interrupted as I was obliged to get Coffee & Tea once more.[67]

For Jacob, it was his little sister's natural place to serve him and his friends, but for Caroline, this expectation began a lifelong irritation with the commands of others, who, as she saw it, sought only to distract her from her ambitions of making a better life for herself.

Isaac Herschel's pride in his youngest daughter was manifested in other ways too. The Herschel children inherited their father's love of music, but, it seems, Caroline and William were also encouraged from an early age in observing the natural world, which intrigued Isaac. Caroline remembered her father's eagerness at showing his children the eclipse of the sun which happened on 1 April 1764, when she was fourteen. Gathering around a tub of water which Isaac had placed in the courtyard, the family witnessed the exciting event in safety. This is mentioned in both of Herschel's autobiographies, so quite clearly it made an impact upon the young girl. Isaac also 'explained the Phenomenon' to the children, who must have been interested, and Caroline sufficiently attentive, in order to remember the eclipse with such accuracy of date and place.[68] Although evidently fearing for the careless dispersal of every penny, Isaac aided William in the pursuit of his philosophical studies and did everything he could to encourage the musical talent of all his children. Isaac's keen interest in astronomy was tellingly passed on to his youngest daughter in a moment which she would recall in old age as immensely important. This was not only a touching event between beloved father and eager daughter, but that daughter's introduction to a field in which she would eventually and successfully participate with her own astronomical discoveries. For Isaac showed his daughter another world when he took her into the street and told her to look up. Herschel was entranced by what she saw: 'some of the most beautiful constellations', and tellingly, a 'Comet'.[69]

Little did Isaac realise when he mapped the heavens for his daughter that, 40 years later, she would be renowned for doing exactly the same.

After leaving compulsory education, Caroline sought to add to what she proudly but modestly called her 'slender stock of self acquired abilities'.[70] But her attempts to improve herself were blocked all the way by Anna, who wanted to keep her unpaid house-keeper; and also by Jacob, who, after Isaac's death in March 1767, just after Caroline's seventeenth birthday, assumed his position as dictatorial head of the Herschel household. If Isaac had lived, Caroline might have been able to pursue the course of instruction he wanted to give her. For Isaac was keen to promote his daughter's chances in the world by giving her 'something like a polished education', which presumably would have allowed her to enhance her talents by learning French, the eighteenth-century *lingua franca*, as well as the language of polish and sophistication, which could open many doors for an impoverished girl without any prospect of marriage in the near future.[71] Jacob and William, as intelligent boys, had received private tuition in French, which would automatically have raised their social status, as the former was only too aware. While tolerating her sons' intellectual elevation, Anna was determined not to allow her daughter to venture too far away from her background. As usual, Anna had an ulterior motive in confining Herschel's studies to the strictly mundane. If she had to learn something, stated Anna, then let it be something which would contribute to the domestic chores. The weak and gentle Isaac was roughly overruled by his vulgar wife and Herschel was permitted only to improve 'useful' elements of her education. As such, she was sent to a seamstress to be taught how to make household linen. This obviously practical task, which could not distinguish her in any way from other girls, almost destroyed Herschel's spirit:

> [H]aving added this accomplishment to my former ingenuities, I never afterwards could find liesure for thinking of anything but to contrive and make for the family in all imaginable forms whatever was wanting, and thus I learned to make bags and sword knots before I knew how to make Caps and furbelows. My destiney seemed now to be unalterable.[72]

Anna's ploy certainly worked. Her daughter was now far too vital to the household even to contemplate dreaming of higher things. To have her youngest daughter working at the most tiresome chores was Anna's only aim in sending her to be educated further. It certainly saved on the expense of a servant, as the elderly Herschel remarked in her autobiography.[73] While Isaac taught his daughter to aim for the stars, Anna thought no further than the scullery or the sewing box.

Anna's disquiet was that of the illiterate who fears, but does not comprehend, the effects of learning. As Herschel put it, generously, considering the dramatic impact her mother's ignorant and misguided policies had upon her life:

> Though I have often felt myself exceedingly at a loss for the want of those few accomplishments of which I was thus by an erroneous though well meant opinion of my Mother's deprived; I could not help thinking but that she had cause for wishing me not to know more than what was necessary for being useful in the family; for it was her certain belief; my brother Wilhelm would have

returned to his Country, and my eldest brother not have looked so high, if they had a little less learning.[74]

But Herschel viewed learning and accomplishment far differently from her mother. For Caroline, they meant achievement and independence. So, with one parent determined that she should not succeed at anything 'impractical' or fancy, she was forced to search for opportunities of improving herself. Like the tactics she had employed when perfecting the art of *stillsitzen*, she had to work around her life of servitude. At the age of sixteen, a chance meeting with a Mademoiselle Karsten, whose parents lived in the same house as the Herschel family, led to one such opportunity. Miss Karsten possessed impressive sewing skills, not just the plain work of repetitive drudgery, which was useful only at home, but the much prized talent of ornamental and fancy work.[75] To have such skills could only advance Herschel's chance of obtaining decent paid employment. Meetings between the two were clandestine (signalled by the consumptive Karsten's loud cough) and had to take place at daybreak, because the unpaid servant was required to start work in her own home at seven o'clock in the morning. This extra tuition was ended abruptly when Miss Karsten succumbed to her disease after about a year.

But Caroline Herschel was not to lose her spirit of independence. After the death of her father and the departure of Jacob, who followed William to find employment in Britain, Herschel 'begged' her mother to allow her to learn a trade – millinery. As the newly appointed head of the household, Jacob dictated that this was acceptable only if his sister was attending instruction 'to learn to make [her] own clothes'. He was 'positively forbiding it for any other purpose'. Jacob's conditions here can only have arisen from the social snobbery of the lower middle classes. No sister of his would become a tradeswoman, earning independence while he ran the Herschel family. Anna's reluctance to send her daughter to be taught millinery was also financial. Extra accomplishments were costly and very highly sought after. For the first time in her life, Anna helped her youngest daughter achieve her aims by persuading the lady who ran the classes to take a lower fee. Herschel not only joined the class with enthusiasm, but found a lifelong friend whom she would later meet again in Britain, when Mrs Beckedorff became lady-in-waiting to George III's consort Queen Charlotte, and William became George's royal astronomer. Although surrounded by several young ladies of more genteel families, who were clearly not attending lessons for any purpose beyond adding to their accomplishments, Herschel managed to gain numerous vital skills in the millinery art.[76] The classes soon came sadly to an end, however, for what, presumably, were financial reasons.

At the age of twenty, Caroline Herschel had increased that little stock of accomplishments, found a friend, visited her cousin in Hameln, and even attended the theatre regularly and without charge, as her brothers played for the orchestra there.[77] She could now proudly list her millinery achievements in both her autobiographical writings and in later letters to her family. A letter to Margaret Herschel from September 1838 provided a portfolio account of her skills: 'for there was no kind of ornamental needlework knotting, platting Hair, stringing beaded Bugles &c of which I did not make samples by way of securing the art.'[78] But, all too often, in the autobiographical accounts, bitterness about the little rewards she received from all her hard work scarred the pride. Angrily addressing her youngest brother Dietrich, who was the original recipient of *The First Auto-biography*, Herschel emphasised how her account should be read by her privileged brother:

[B]y what is to follow D[ietrich] may also see how vainly his poor Sister has been strugling through her whole life, for acquiring a little knowledge and a few accomplishments; as might have saved her from wasting her time in the performance of such drudgeries and laborious works as her good Father never intended to see her grow up for.[79]

Tellingly, this outburst came straight after Herschel's burning of the account of her early life. Exploited and enslaved, she expressed her fury at the treatment her family meted out to her; but, as her father had anticipated, she had her eyes on higher prospects.

Yet Herschel remained confined to the household, unable to employ her newly-found achievements to any worthwhile purpose. The terror of being forced into servitude, losing her identity as an 'Abigail or Housmaid'[80] outside the home, seemed to keep her just where Anna wanted her. Without the vital knowledge of French, she could not take a place as a governess, and family pressures forbade her from employing her millinery skills in the world. Terrified of 'dependance', Herschel was at the end of her tether.[81] Summing up the first 22 years of her life in Hanover, at the age of 77, she stated succinctly in a letter to her nephew that she 'had been sacrificed to the service of my family under the utmost self privation without the least prospect, or hope of future reward'.[82] No matter how hard she tried, her chance of a stable future was removed ever further from her grasp. But William Herschel, who had moved to Britain and settled in Bath in 1766, was about to offer his sister the opportunity of a lifetime, a chance to fulfil her ambitions and earn her independence. At the age of 22, Caroline Herschel's life was about to take an unexpected turn.

Chapter II

Escape to Bath

Since December 1766, William Herschel had been living in one of eighteenth-century Britain's most fashionable and bustling towns: Bath. Within a short time he had made a name for himself as a musician, becoming the organist at the newly established Octagon Chapel and a sought-after music teacher. Over the next few years, he found temporary situations for his brothers, Jacob, Alexander and Dietrich. Now, it was the turn of his younger sister, whose potentially limited period of employment with her brother would not only turn into 50 years of devoted service, but also allow an ambitious woman to begin to realise her long desired aims of independence and personal and financial security.

In her autobiographies, Caroline Herschel suggested that by the end of 1771 she was becoming desperate about her future prospects, terrified of 'becoming a burden to [her] brothers' and convinced that she had no hope of ever marrying or escaping her life as a Herschel family domestic slave.[1] At the age of 21, she had managed to wring some educational concessions out of her reluctant mother and elder brother, but she was still painfully aware of her deficiencies in skills which would matter outside the home. In October 1771, however, William wrote from Bath to make a suggestion to his younger sister which would offer her a way out of her apparently eternal, dismal situation. Alexander had told William that Caroline 'had some Notion of Müsic and a good Voice'.[2] William proposed, therefore, to return to Hanover the following summer to collect his sister and 'make a tryal' of her musical abilities.[3]

For a young woman as dedicated to improving herself as Caroline Herschel, this opportunity both to escape Hanover and the oppressive Herschel household and to make a career for herself doing something she enjoyed, must have presented an enormous and exciting challenge. While, with hindsight, she was able to deal with the event calmly, the momentous letter from William must have been one of the most important moments in Herschel's life. Certainly, the indication that she 'set [her] heart upon this change' reveals the wealth of potential to be gained from a move to Britain and the deep, even physical, pride she felt in being valued enough in the first place to be 'tried' on her ability to adapt herself to new circumstances.[4] In both autobiographies, indeed, she employed repeatedly one of her favourite terms: 'useful'.[5] To be useful to others, while also using her talents to boost her own ambitions, self-esteem and independence, would become Caroline Herschel's dual *modus vivendi*. As she wrote in a letter to her nephew's wife, 65 years later, the invitation to Bath allowed her finally to set in motion long-cherished hopes and a life dedicated to achieving her goals. The first 22 years of her existence had been scarred by isolation and oppression, but the next would be characterised by events memorable for all the right reasons. 'I most [must] think no more of those times', claimed the elderly Herschel, surveying her Hanoverian past, 'only just say I came to Bath with a mind eager to learn and to work, and never changed any mind till I came here again.'[6] Unlike her overreaching, skittish brothers, finally Caroline Herschel would have made her father proud.

But first Herschel had to pass the 'tryal', and in order to do so, she had to train herself to become a member of a profession in which the majority of her family was engaged. The life of a musician, as Herschel was aware, was far from comfortable. In eighteenth-century Europe, music was 'a profession which embraced extremes of fame and obscurity, genius and mediocrity, mobility and quiescence'.[7] With all her brothers and her father choosing to follow their musical talents for a living, Herschel had seen at first hand the often itinerant, peripatetic, financially insecure lot of a dedicated musician. This situation was not unusual in the fragmented states which made up what is now known as Germany. Due to the diffusion of culture and commerce in northern Germany, travel was fundamental to the development of both. Musicians, teachers and opera companies were required to live a nomadic lifestyle, but artistic tours could lead to employment as a private tutor or in a Kapelle, a court's musical household.[8] With the absence of the monarch, who was simultaneously, from 1714, King of Great Britain and Elector of Hanover, Isaac and his sons had chosen to pursue their careers in a state without a central and cohesive court presence like much of the rest of Germany. Without an active and lively court or a monarch to lavish enormous sums in order to develop music within the electorate, opportunities were few and far between. No wonder the Herschel family was forced to move around and even to leave the country in order to improve the prospect of success in their musical careers.

Yet at least Jacob, William, Alexander and Dietrich had been encouraged to practise at home, as well as to reveal their talents in public arenas. So far, the only time their sister had seized the chance to play was after her father had finished teaching Dietrich or his other pupils, and even then only when her disapproving mother was absent. Herschel's enjoyment of and desire to improve her performance on the violin characterises the first part of her autobiographies, and she reveals again and again her resentment at the time spent away from her practice. At the age of eighteen, Herschel found her time taken up with entertaining her cousin. For a girl so bereft of friends, her reaction to her cousin's attentions might seem surprising:

This young Woman, ful of good nature and ignorance grew unfortunately so fond of me that she

was for ever at my side, and by that means I lost
that little interval of leisure I might then have
had for reading, practising the Violin, &c entirely.[9]

But a chance to improve herself could preclude even
the possibility of friendship for the lonely girl. Constant chatting about trivialities was a disastrous dis-
traction when one could be working to enhance that
little stock of accomplishments.

So, when William contacted his family to request
a musical 'tryal' for his sister it was with some misgivings. How could a young woman, who was rather
unsociable and used to picking up the violin only
when no one noticed, become a professional musician or sing in front of large, public audiences at
concerts and oratorios? William was undoubtedly
taking a great gamble when he proposed the move to
his family. But, although such a change would require
innate ability, which had yet to be fully unlocked for
various reasons in Caroline's case, it also necessitated
intense hard work, courage and persistence, qualities
that William must have known his sister possessed
in abundance. This faith in Herschel's latent abilities,
however, did not extend to the more doubtful Jacob,

whom William had asked to start instructing his sister before he arrived. Jacob, in similar fashion to his mother, treated his sister like a servant and had no intention of helping her; in fact he found the whole scheme hilarious. Herschel's frustration with her amused and unhelpful brother was apparent in both versions of her autobiography. Instead of 'cultivating' his sister's voice, Jacob 'made him[se]lf very merry, at [her] thinking seriously about – on what he chose to turn into a joke'; the 'whole scheeme' was turned into 'ridicule'.[10] Caroline Herschel was to be abandoned, and not for the first time, to help herself.

And in typical Herschel fashion, she did. Jacob's behaviour had succeeded only in encouraging his stubbornly persistent sister to prove him wrong. Caroline's 'harrassing uncertainty'[11] stemmed from not knowing whether she was allowed to go to Bath, rather than from any doubting of her own ability to accomplish the task her brother William had set her. If anything, then, Jacob's mockery had precisely the opposite effect to that which he had intended. His sister was now more determined than ever. In order to train her voice, Herschel waited until the house was empty. With her negative brother and disapproving mother absent, she could teach herself how to sing without any disruption. And she made sure that she was occupied with a household task while she sang, in order to deflect the scorn or criticism of prying friends and neighbours, indignant that she was neglecting her domestic duties. In her *Second Autobiography*, Herschel noted proudly that for this practice she

> found opportunity enough, for my Brothers were seldom at home, and my Mother frequently spending the afternoon in visiting some Friends; so that when I found myself alone seated in a back-room, with closed windows; I could Ply my needle and exercise the Voice without molesting the Ears of any one.[12]

Knowledge of the violin allowed Herschel to adopt an approach which helped her with her singing voice. Scrupulously professional, she wanted to give herself the best start, and so followed the 'rules given to beginning Scholars by singing with a Gag between their front Theet [teeth]'. This enabled her to imitate, impressively from memory, the 'solo parts of Concertos shake and all such as I had heard them play on the Violin'. Herschel's method ensured that she 'gained a tolerable execution' even before she 'knew how to sing'.[13]

Having begun to train her voice, Herschel continued to appease the relatives she would leave behind. If she saw herself as somehow 'deserting' her family, she certainly made sure that they could not recall her for domestic duties. Herschel planned her departure with military precision. Her voice came first, but then, deliberately, she made herself useful for Jacob and Anna, the two most demanding and disapproving members of her family, and for her younger brother Dietrich, now sixteen:

> I next began to nit ruffles, which were intended for my brother W[illia]m – else they were Jacobs. For my Mother and brother D[ietrich] I knitted as many Cotton stockings as were to last two years at least. In this manner I tried to still the compunction I felt at leaving relatives who I feared would lose some of their comforts by my desertion; and nothing but the belief of returning to them full of knowledge and accomplishments could have supported me in the parting moment.[14]

It is noticeable that Herschel worried for the loss of

her family's 'comforts', not of their necessities. As an unpaid servant in the household, Herschel, after all, spoilt her family members when she waited upon their every whim; a fact of which, here, she was clearly aware. The loss they will feel is nothing, she suggested, compared to the necessary and worthwhile development of her skills, which should make them proud of her achievements. And if she did not succeed in Bath as a singer in William's concerts, at least her family would not want for stockings in the two-year 'tryal' period she had been allotted.

William Herschel arrived, as promised in his letter of the winter before, in August 1772. As head of the household after the death of Isaac, Jacob had to be applied to for his sister's removal, but as he was away working as part of the entertainment for the Queen of Denmark some distance away at a royal hunting lodge, Göhrde, near the Elbe, William and Caroline left without his formal consent. Caroline suggested in *The Second Autobiography* that, in removing his sister from her vital domestic position, William actually compensated his mother for the loss of her unpaid help. As Caroline so succinctly put it: 'the anguish at my leaving her being somewhat aleviated by my

Brothers settling a small Annuity on Her, by which she would be enabled to keep a person for supplying my place.'[15] For Anna, the loss of her daughter was not a wrench in that she would miss her company, but because her presence was vital in order to keep the household running smoothly. As far as Anna was concerned, the employment of a servant eased her youngest daughter's departure. Thus, Caroline and William Herschel could leave for Bath on 16 August 1772, safe in the knowledge that the former could be replaced easily and that there were enough stockings to last until she returned.

Caroline had escaped her hellish Hanoverian existence, but now she was forced into another maelstrom – an entirely unknown country, and one which spoke a language of which she knew not one word. In fact, they nearly did not arrive at their destination at all. The terrible storm which accompanied their crossing from Holland was not an auspicious start, and the all-but-wrecked ship limped into the country which was to be Herschel's new home having lost its main mast.[16] Another disaster befell the party after landing in Harwich. When travelling to meet the public stage coach which would take them to London, the horse pulling the cart, unused to being harnessed in this way, bolted and overturned the vehicle, spilling the passengers into nearby ditches. Luckily, no one was injured and the journey continued. It would have been little wonder if even the sensible, never superstitious Caroline Herschel had felt as if fate was catching up with her for making such a momentous decision to leave her home and change her life.[17]

Eventually arriving in London, the Herschel siblings managed to find the time to visit a few metropolitan monuments such as St Paul's. It was quite a frenetic tour, as Herschel claimed that the only time they were 'from [their] legs' was to eat their meals. However, like another German visitor, Sophie von La Roche, fourteen years later, Caroline Herschel noticed more when the shops were, as she put it, 'lightet up' in the evening.[18] This indication that she was about to enter the very different world of display and conspicuous consumption which characterised eighteenth-century Britain may have prepared her for her new home, the fashionable city of Bath. The fact that they lingered most longingly and most memorably around the windows of optician's shops added another dimension to William's new life – his 'growing obsession with astronomy', which would deprive his sister of the promised assistance with her singing. Perhaps another 'ill omen' to add to the growing list since they had left Hanover.[19]

The Herschels arrived, exhausted, in the city on 27 August 1772. Bath is now known and marketed as a quintessentially Georgian city, but even contemporaries recognised elements of the world which surrounded them displayed in miniature. Bath offered a curious combination of nature and culture, its natural springs exploited and the city converted into a profit-making spa resort in the eighteenth century. Bath's intended status as a place for the convalescence of invalids was thus undermined with the growth in the variety of entertainments on offer for visitors and their more healthy relatives. The city was crowded with a heady mixture of those who took the waters and those who were interested in the other liquids on offer in the numerous taverns. It also became notorious for rakes searching for eligible heiresses and less than eligible ladies who indulged in a spot of wealthy husband-hunting. The number of sick people also encouraged the presence of quacks, who dispensed advice and medicine to the ill and desperate. For those who sought a cure for illness and for those who sought only pleasure, Bath was an ideal location.[20] Published a year before Caroline Herschel joined her brother in Bath, Tobias Smollett's epistolary novel *The Expedition of Humphry Clinker* satirised the city as a microcosm of all that was rotten in contemporary British society. Accompanied by his sister, niece and nephew, Smollett's curmudgeonly main character, Welsh squire Matt Bramble, griped that Bath encouraged the horrible bustle he so despised, and that it displayed the supposedly sophisticated world as an absolute fraud. For Matt, however, it was the British obsession with finance which had caused ultimate chaos, encouraging even the lowliest to buy their way to a higher station. This social elevation was on show in Bath more than anywhere else. As he expostulated:

All these absurdities arise from the general tide of luxury, which hath overspread the nation, and swept away all, even the very dregs of the people. Every upstart of fortune, harnessed in the trappings of the mode, presents himself at Bath, as in the very focus of observation – Clerks and factors from the East Indies, loaded with the spoil of plundered provinces; planters, negro-drivers, and hucksters from our American plantations, enriched they know not how; agents, commissaries, and contractors, who have fattened, in two successive wars, on the blood of the nation; usurers, brokers, and jobbers of every kind; men of low birth, and no breeding, have found themselves suddenly translated into a state of affluence, unknown to former ages; and no wonder that their brains should be intoxicated with pride, vanity, and presumption. Knowing no other criterion of greatness, but the ostentation of wealth, they discharge their affluence without taste or conduct.[21]

While Matt criticised the lower classes for their misdirected and ostentatious displays of wealth, his greatest hatred was for the shady origins of such unnecessary consumerism; for financial gain achieved from the 'spoils of plundered provinces' and the use of inhuman slave labour. Matt's disgust was focused on the subsequent feeding of this gain back into the British economy, so that the country was effectively run upon exploitation and inhumanity. Caroline

Herschel was used to servitude, but would her life as a singer in Bath allow her to escape from her lowly position, or would she, as Matt Bramble would have put it, allow herself to become enslaved by the unhealthy air of this most fashionable of eighteenth-century spa towns?

For Bath had a third side to its structure of pursuits. Alongside illness and entertainment was the livelihood of those who provided amusement and existed to please the convalescent. As well as a revival in the city's fortunes, the eighteenth century witnessed the apex of opportunity for the musician in Bath. In order to entertain the ever increasing number of visitors, Master of Ceremonies and social organiser *extraordinaire* Beau Nash paid half a dozen musicians a guinea a week to play for the crowds. With the growth of visitors came the inevitable increase in sociable public spaces like the Pump Room, which was popular in the morning for strolling and taking the waters, and the Assembly Rooms, where the evening entertainment could be found. Seven instrumentalists were hired, for two guineas, to entertain the audiences at each venue. Although 'in origin a small undistinguished rural town, Bath was forced to acquire many of the institutions usually confined to a metropolis'.[22] Of course, with all the benefits came all the ills that such a dramatic alteration would bring. Tobias Smollett's Matt Bramble certainly captured the fashionably idle nature of those who spent the 'season' in this most popular of spa towns. It has been estimated that in 1750, Bath had an indigenous population of around 10,000, but had the facilities in place to cope with an influx of 12,000 or more during the season, which lasted from October to Easter. By 1801, the population had risen to 33,000 and 147 coaches from London arrived every week.[23] But, despite the seemingly endless opportunities for musical employment, life was a great deal more complicated for the profession than the statistics might suggest. For a start, the ratio of visiting population to musician was dramatically disproportionate. While London could boast maybe 1,500 musicians in the mid-eighteenth century, only the university cities of Oxford and Cambridge, Dublin, Bath and, briefly, Edinburgh could employ around 50 full-time musical practitioners.[24] In some ways, life had become more unstable for the profession. The increasing commercialisation of music allowed new freedoms for the musician to perform in front of diverse audiences. But this release from performing for small, intimate, cultivated and demanding circles to relying on less discriminating, less reliable pupils and audiences posed new risks for the unwary.[25] While the potential audience had grown, due to the movement away from court entertainment since the Restoration in Britain, the possibilities for musical employment had become far more precarious, far more uncertain. Theatres and subscription concerts could provide relatively stable jobs, while provincial festivals offered work for only a few days. Private engagements and music lessons could prove a profitable sideline. But, on the whole, the development of opportunities for the profession was 'slow [and] cumulative' and always seasonal, subject to endless fluctuations of taste, economics and politics.[26]

When Caroline Herschel joined her brother in Bath in an attempt to mould herself into a singer, William had already built up quite a successful business for himself in the city. He had already experienced the precarious nature of the musician's peripatetic life, both in Hanover and in Britain, where he had failed to earn a living in the early 1760s copying music in London and had later joined the York-shire militia band.[27] Eleven years before William Herschel brought his younger sister to Bath, he wrote of the frustrations of the piecemeal income to be obtained from music:

> You don't perhaps know that I have already some time been thinking of leaving off professing Musick and the first opportunity that offers I shall really do so. It is very well, in your way, when one has a fixed salary, but to take so much for a Concert, so much for teaching, and so much for a Benefit is what I do not like at all, and rather than go in that way I would take any opportunity of leaving off Musick; not that I intend to forget it, for it should always be my chief study tho' I had another employment. But Musick ought not to be treated in that mercenary footing.[28]

Supposedly an art which provided pleasurable entertainment, music became, as William so astutely realised, a trade to be pursued solely for profit in the second half of the eighteenth century. But, by 1767, his position as organist in the Octagon Chapel in Bath allowed William to begin making his name in the music profession by employing the tactics he had previously despised in order to succeed among the intense competition. The Octagon opened officially on 4 October 1767 and remained in regular use until 1895. An influx of visitors to Bath during the 1750s and 1760s had encouraged the increased building of both public entertainments and private residences to accommodate domestic and social needs. Religious obligations were also catered for with the development of several privately-owned chapels, which provided a 'comfortable, almost domestic, setting for religious exercises now come to be treated as part of the social round'.[29] The Octagon was one of the most successful of these new establishments, built with funds from a subscription organised by an amalgamation of religion and finance in the form of the Reverend Dr John de Chair and the banker, William Street.[30] William Herschel played the organ on its inauguration on 28 and 29 October 1767, for two performances of Handel's *Messiah*, held for the 'relief of the industrious poor'. The opening of the chapel was also an opportunity to advertise his musical ability.[31] Within ten years, the Octagon would also witness the debut of Caroline Herschel in her new career.

But all this was to take place some time in the future. When Caroline and William arrived in Bath on 27 August 1772 they went straight to the latter's lodgings at 7 New King Street, where Alexander, who had been living with William, would later join them. As an illustration of the precariousness of the music profession, Alex Herschel had been forced, when the Bath season ended at Easter, to go and look for another engagement, travelling down to Southampton to find work. The house was shared with the Bulman family, with whom William had boarded when employed by the Yorkshire militia in Leeds. Mr Bulman's financial collapse had sent the family to Bath, where William had kindly found Bulman a job as clerk to the Octagon.[32] The realisation of the incredible upheaval she had effected in her life now hit Caroline Herschel particularly hard. Her alien status was immediately apparent, both to herself and to others: 'And so I found myself all at once in a strange Country and amongst straingers.'[33] Now Herschel finally had the opportunity – or so she thought and expected – to perfect her musical talents in readiness for her first performance as a professional singer.

But there were other more immediate tasks to

accomplish first. Herschel had arrived in Britain not knowing a word of the English language, and could only repeat what she was told, 'like a Parrot', she recounted in old age.[34] Not only would she be required to sing in the language when the time came, but, in order to manage the Herschel household – for which task she found herself, typically, liable – she needed to be able to associate with tradesmen, as well as guests. This mingling with shopkeepers was something that not even Anna had made her daughter carry out, and the prospect terrified her. After only six weeks and some intensive English lessons from her brother, Caroline Herschel was released into the streets of Bath to gather provisions for herself and her brother. Herschel narrated the story in both autobiographies, noting how vehemently her 'particular dislike' was for 'Shopping and Marketing':

> Sundays I received a sum for the weekly expenses of which my Houskeeping book (written in English) shewed the sum laid out and my purse the remaining Cash. One of the principle things required, was to market myself, and about six weeks after my being in England I was sent out

alone among Fishwomen, Butchers basket women &c and brought home whatever in my fright I could pick up.

Unbeknown to Herschel, Alexander, who had returned recently from his summer engagement, watched his sister from afar in order to secure her return.[35] With only a smattering of English, this experience must have been terrifying for Caroline, who ultimately felt that frightened snobbery for trading which she so despised in Jacob.

But English was not the only lesson Herschel was learning in Bath. Although the Herschels employed one female servant, some of the old Hanoverian drudgery fell upon Caroline. With some bitterness, she complained that 'The making, and keeping the Linen in repair fell also to my share for there was no one else who would do it'.[36] When she had been confined to the household in Hanover, Herschel had been required to complete all the lowly duties which her mother passed on to her daughter, but in Bath, she was placed in the situation of housekeeper and was encouraged to learn accordingly. This was often the lot of the spinster sister who looked after her relatives. It was a fairly common occurrence to find unmarried women performing the role of unpaid housekeepers to their bachelor brothers, but they also acted as servants within their own family: 'The upper ranks of the domestic service hierarchy offered one of the few openings available to those who needed to earn a roof over their heads but wished to remain acceptable to genteel society.'[37] Caroline Herschel had moved rapidly up the ranks of domestic servants from scullery maid to the more genteel role of housekeeper in one large leap. Consequently, Mrs Bulman was instructed by William to teach his sister various forms of housekeeping:

I began to think on how those hours I should be left to myself might best be spent in lerning what would be necessary to know for a houskeeper of our little Family; and for this purpose the first hours imediately after breakfast were spent in the Kittschen, whe[re] Mrs Bulman taught me to make all sorts of Pudings and Pies, besides many things in the Confectionary business, Pickling and preserving &c &c – A knoledge for which it was not likely I should ever have any occasion.[38]

Herschel's attitude to her new accomplishments is rather curious and certainly ambiguous. 'The Art of Preserving, and Candying, Fruits and Flowers, and making all sorts of Conserves, Syrups, Jellies and Pickles' was to be expected of the 'accomplished' eighteenth-century female wishing to run a household.[39] However, such culinary skills Herschel clearly felt were below her. She should have 'no occasion' for these domestic talents, either because she would never have her own household, having been instructed from an early age about her lack of marriageable qualities, or because she now had a prospective career, which would take up all her time. Considering Herschel's ambitious attitude, it seems as if the latter might have been the most obvious explanation for her reluctance to learn skills which were alien to her musical development.

Alongside instruction in these more womanly duties, Herschel was being tutored by her brother in arithmetic in order to keep the household finances running smoothly. At breakfast, therefore, William taught his sister the English language and basic accounting skills. To begin with, William was at 'leisure' to help his sister, giving her '2 or 3 Lessons every day'; the mornings were spent at the harpsichord where William continued to develop his sister's voice through practice with the 'gag' method by which she had already taught herself. She noted proudly that her attempts at self-education had been extremely successful, as William was 'very well satisfyed' with her ability. Herschel's dogged persistence, in spite of the meanness and mockery of her elder brother Jacob and the disapproval of her mother, had clearly paid off. And yet, 'managing the family' and learning English seemed to take up most of the time that Herschel felt could have been spent on more vital singing practice.[40] The more the list of tasks grew, the more worried Herschel became about her beckoning career, hence her lack of interest in perfecting even the most simple culinary arts. But even before the season began again, Herschel was abandoned frequently by her brothers, who did not have the time to superintend their sister's much desired vocal progress. Again, Herschel was left to improve herself, 'with directions how to spend 5, 6 &c hours at the Harpsichord', but without anyone except herself to applaud her achievements. Paying customers would always take precedence for William Herschel over his desperately eager younger sister.

Caroline Herschel's existence during the first few months in Bath must have been isolated and depressing. So much so, indeed, that she even found herself succumbing to 'Himwehe' or homesickness and 'Lowspirits'.[41] Even life as Anna's maid appeared preferable to the loneliness of a foreign land, where one was doubly separated from the world through barriers of language and lack of acquaintances. Herschel 'knew too little English for deriving any consolation from the society of those who were about [her], so that during dinner time excepted, [she] was left intirely to herself'. Despite all its promises, Bath presented Herschel with nothing more than a mirror image of Hanover at this time, except that in the former, she was surrounded by familial bustle. Alex most certainly did not help:

> For my brother Alex. was also much engaged with public business and giving Lesson on the Violoncello and if at any time he found me alone, it did me no good, for he never was of a cheerful disposition but always looking on the dark side of every thing, and I was much disheartened by his declaring it to be impossible for my Brother to teach me

anything which would answer any other purpose but that of making me miserable.[42]

The musical life, suggested Alex, the trials of public performance for little reward, could by no means answer the desires or ambitions of his sister. She had better, he implied, never have agreed to their brother's scheme and remained unfulfilled as an unpaid servant in her own home in Hanover.

And yet Caroline refused to be bowed down either by her own doubts or by the misgivings or pressures of others. Her devotion to her new musical pursuits was signalled by the frustrations experienced when dealing with a succession of disruptive domestics who disturbed Herschel's practice and ensured that she had to devote herself solely to looking after the household.[43] Now she had more important things with which to occupy herself, she was not satisfied with domestic duties. Nor was she comfortable with the female company she was encouraged to keep. Desperate for a friend in whom to confide her fears and worries, Herschel was not allowed to form an acquaintance with anyone 'but such as were agreeable to my eldest Brother'.[44] Even at this early stage in their long life together, Caroline experienced frustration with the straitjacket that her mostly beloved brother imposed upon her movements. With such restricted access to company, loneliness, she decided, was preferable to the interference and gossip of trivial women. Although William had suggested from the moment they arrived in Bath that Bulman's wife and daughter would provide ideal and close company for his sister, as they were already on the premises, Herschel found them and other recommended potential company appallingly stupid and disruptive:

> Mrs _____ and her Daughter were very civel and the latter came sometimes to see me, but being more an[noyed] than entertained by her visits I did not encourage them, for I thought her very little better than an Idiot. The same opinion I had of Miss Bulman, for which reason I could never be sochiable with her.[45]

Herschel's dislike of female trivialities and her abrupt manner are everywhere apparent in the autobiographies. The fact that she did not believe in expressing civility to those who did not deserve it was made clear when William was compelled to provide his sister with a list of social niceties. In order to obtain the 'good graces' of potential supporters and patrons, the 'very great Critickers on Singers and Müsical performers', such as the wife of the Reverend de Chair who had set up the subscription for the Octagon Chapel, William 'laid' 'a few of those flattering Com[plimen]^{ts}' in his stubborn sister's mouth.[46] Caroline would please only by her actions, only by the strength of her voice, not by her mastery of dictated airs and graces.

Another woman, Mrs Colnbrook, was engaged by William to smooth out his sister's rough ways by taking her for an extended six-week stay in London in January and February 1774, which would be Caroline's first and only proper introduction to the metropolitan social whirl. Herschel's second glimpse of the British metropolis was to involve more lights than those she had noticed illuminating the shops at night on her first visit. To make an entrance into the world was a momentous event for a young lady. The novelist Frances Burney, who would meet the Herschels a decade later, described the London of the 1770s in her first work, *Evelina* (1778). As Burney describes it, maintaining a 'good' behaviour in public was exceptionally difficult for any young woman whose every move was scrutinised by her elders, contemporaries and potential suitors. Coupled with the uncomfortable surveillance was an endless round of sociable entertainments, involving dressing and undressing, manners and etiquette, causing the eponymous heroine, who had been brought up in the countryside, to exclaim: 'I think there ought to be a book, of the laws and customs *à-la-mode*, presented to all young people upon their first introduction into public company.'[47] Although typically spiky about being 'every night pushed about in a gentil crowd', and understandably concerned about every penny she spent, Herschel actually seemed to conduct herself satisfactorily and even to enjoy herself. In her *First Autobiography*, she remembered the hassles the trip caused her. Her account was peppered liberally with 'suffering', 'dread', and 'expense'.[48] The second narrative of the events was more detailed, and as well as remarking upon two visits to the Pantheon, that 'wonderful and beautiful place', Herschel expressed disappointment at missing the famous actor David Garrick's final stage performance.[49] Perhaps sensing that this trip would have pleased her father's desire for her to acquire some 'polish', Herschel treated it as part of her education, whereby she could learn not only from the singers at the opera, but from the behaviour of the ladies in the audience.

By 1774, Caroline Herschel was pleased with an improvement in the range of her voice. Although her brother had set out directions for her instruction, Herschel had put the theory into practice while she was alone during the long days when her brothers were working. But her frustrations were mounting again. William had grown increasingly obsessed with his new interest – astronomy – and was putting less energy not only into his own music, but into his sister's training, and, of course, she had still not made her first public appearance. Instead, he was now keen to reverse the assistance, and wanted her to help him to follow his latest pursuit. Not content 'with what former observers had seen', William Herschel 'began to contrive a telescope of 18 or 20 feet long'. Caroline was pressed into assisting, and as she puts it unhappily: 'I was to amuse myself with making the tube of pasteboard against the glasses arriving from London.'[50] Further disruption occurred when the Herschel home became a 'workshop', and then, in the summer of 1774, the family were compelled to move slightly further away from the centre of Bath to Walcot, to a house which allowed more room for William's telescope construction and a 'place on the roof for observing'.[51] He was either too exhausted or too busy to help his sister, and Caroline was once again abandoned. Indeed, she was soon to be reduced to her brother's nursemaid. In a famous passage which has been frequently employed to illustrate Caroline's selfless devotion to the interests of her brother, to the detriment of her own life, Herschel suggested that William was so absorbed in his activities that

> by way of keeping him alife I was even obliged to feed him by putting the Vitals by bitts into his mouth – this was once the case when at the finishing of a 7 feet mirror he had not left his hands from it for 16 hours together. … And generally I was obliged to read to him when at some work which required no thinking, and sometimes lending a hand.[52]

Housekeeper, unpaid apprentice, and now nanny; Herschel's expectation that she was to sing for her long-cherished hope of gaining an independent

income was being mocked every day that she continued to serve her brother.

Caroline Herschel's participation in astronomical
activities would have been viewed ambiguously
by contemporaries. The mechanics of the heavens
attracted surprising numbers of women, encouraging
an astronomical craze at this time. As the nineteenth-
century French literary critics, the de Goncourt
brothers, noted wryly in a book on eighteenth-century
women: 'Novels disappeared from ... dressing-tables
...; only treatises of physics and chemistry appeared
on their *chiffonières*.'[53] Partly, in Britain, this was due
to national pride in Newtonian physics, and this
patriotism was reflected in the number of publications which sought to explain Newton's complex
theories to an eager though bemused audience.[54]
British women were particularly prominent in the
translation of Newtonian distillations for the popular
market. The playwright Aphra Behn translated
Fontenelle's *Plurality of Worlds* in 1688, while Elizabeth
Carter's version of Algarotti's *Newtonianism for the
Ladies* was reprinted three times between the 1740s
and 1760s. Both texts featured the dialogue format,
and scientific exploration thus appeared as an invig-

orating, sociable activity rather than the solitary occupation implied by stargazing.

However, there was also potential female unease about the pursuit of scientific studies revealed in the translations of Behn and Carter. Unlike, for example, the securely domestic scene of Benjamin Martin's *The Young Gentlemen's and Ladies Philosophy* (1759–63), in which a brother passes on his expertise to his sister within the confines of their own home,[55] the translated texts are dependent upon a transfer of knowledge from man to woman in public spaces. The latter introduced ideas in celestial mechanics through flirtatious dialogue and romanticised, fictional situations. The male philosopher in Behn's translation of Fontenelle's *Plurality of Worlds* notes rather slyly that: 'Mathematical reasoning is in some things near a-kin to Love; and you cannot allow the smallest Favour to a lover, but he will soon perswade you to yield another, and after that a little more, and in the end prevails entirely.'[56] It may not be so surprising that Elizabeth Carter did not value her translation of Algarotti very highly, despite believing in the benefits of an expanded female education. For her, scientific discoveries were too often the result of haphazard showmanship.[57]

But others argued that stargazing in fact brought one closer to the wonder of creation and thus remained firmly within the framework of a religious and moral female education. The fascination with mid-nineteenth-century evolutionary theory has often obscured the evidence to suggest that science and religion were in no way mutually exclusive in the eyes of many before and even after this time. Indeed, analyses of earthly and celestial mechanics proved rather than questioned the existence of a benevolent deity. Reducing celestial activity to mathematical formulae and later placing the world under a microscope, the scientific expositor and practitioner Mary Somerville found similar comfort in her findings of the beautiful magnitude of creation. Somerville was catapulted to fame with her translation of the notoriously complex treatise *Méchanique Céleste* of Pierre-Simon Laplace, published between 1799 and 1828. Somerville's own 'Preliminary Dissertation' to the translation of 1831 resonates with the sublimity of the scientific enterprise:

The contemplation of the works of creation elevates the mind to the admiration of whatever is great and noble, accomplishing the object of all study, which ... is to goodness, the highest beauty, and of that supreme and eternal mind, which contains all truth and wisdom, all beauty and goodness.[58]

Rather than distancing oneself from the deity through the rigour and scrupulosity of science, one can only move closer to God in recognition of the nobility of creation.

Although suggesting that her actions were nothing but the performance of the most mundane, if extreme tasks, Herschel also drew attention to the professional nature of her assistance. Although directing William's son John to the poverty of her sisterly aid in her *First Autobiography*, she said indeed 'so much of' herself that it was clear she viewed her contribution to the construction of telescopes as vital. Why else would she employ the phrase 'I became in time as useful a member of the workshop as a boy might be to his master in the first year of his apprenticeship'?[59] Herschel had certainly undergone an apprenticeship in household drudgery with her mother and was now embarking upon another lengthy period of learning with her brother. In the eighteenth century, '[m]ale apprentices typically spent seven years learning a craft or trade with a master or mistress, though much of their time, especially in the early years, was spent on menial tasks. Female apprentices did exist in this period, to be found in a range of "male" trades such as carpentry, blacksmithing, ironmongery, and bricklaying, but their apprenticeships lasted for around four years.'[60] More common were women who were trained in household duties, as Herschel had been in Hanover. Such training, of course, implied a lifelong apprenticeship, which would never result in properly paid employment. Caroline Herschel was clearly becoming concerned that her apprenticeships never seemed to lead to financial independence and the comfort and security this would bring.

Herschel's reduction to carrying, fetching and even feeding her brother was ultimately to her 'sorrow'; attention was diverted from her own musical education, and she was 'much hindered' in her practice by attempting to learn another new trade. The switch from William aiding Caroline in her career to Caroline assisting William in the development of his new interest provoked the elderly Herschel to an angry paragraph in her *First Autobiography*:

> And in this place I will remark that very seldom have I been so fortunate to meet with gratitude or good will in any I have had to deal with, and many times I could not help thinking but that it was owing to a natural antipaty the lower class of the English have against forigners. In short I have been thoroughly annoyed and hindered in my endeavours at perfecting myself in any branch of knowledge by which I could hope to gain a creditable livelihood; on account of continual interruption in my practice by being obliged to keep order in a family on which I was myself an dependent.[61]

Although ostensibly an attack upon the prejudices of English servants, who continually destroyed Herschel's practising time, there was another target here. The *volte face* by her brother was seen by Herschel as a contributing factor to the destruction of her ambitions. William, like Anna, would allow his sister to achieve only so much. An attempt to gain independence could so easily be thwarted by family members keen to keep their unpaid help eternally dependent.

This continual irritation at her dependency was lifted slightly by the prospect of Caroline's impending first public appearance. In 1775 and 1776, as well as helping her brother with his astronomical activities, Herschel was preparing, when she had the chance, for her performance. As 'Musick was seldom thought of' in the Herschel household now, she was once again left to practise alone.[62] Sometimes she accompanied William when he visited his pupils, and joined in singing lessons or assisted with the Octagon Chapel choir. She began copying music in 1776, as well as preparing the chorus singers for the following Easter's performances of Handel's music:

> I set about copying from the Scores of the Messia and Judas Machabeus into parts for an Orchestra of nearly 100 performers against lent 1777 and the Vocal parts of Samson; the Instrumental parts I was obliged to leave to other hands, for much time was taken up by frequent rehearsals of the Chorus singers of which the teaching of the Treble was left to me.[63]

This breathlessly unpunctuated account of her activities in the latter half of 1776 certainly gave an indication of how frenetic Caroline Herschel's life had become. But some time was to be devoted to increasing Herschel's stage presence, and she was 'drilled' by a famous dancing mistress, Miss Anne Fleming, who was to polish her pupil and make her appear a 'Gentlewoman'. 'God knows how she succeeded', exclaimed Herschel in her *First Autobiography*, but evidently she did.[64] In appearance too, Herschel was every inch the professional. William had parted with ten guineas for appropriate clothing so that his sister could appear to greatest advantage in public for the first time. Despite years spent making clothes for her family and herself, Herschel seemed to know just what was suitable, and clearly had learnt from her experience of mixing with the *bon ton* in Bath and in London. Her father would have been incredibly proud to hear the comment made by Mr Palmer, proprietor of the Bath Theatre, that his daughter was 'an Ornament to the Stage'.[65] Now a trained and polished professional singer, 26-year-old Caroline Herschel, former Cinderella, current budding astronomer's apprentice, was ready for her public debut.

During Lent 1777, Herschel sang for the first time

as a principal in the oratorios she had copied out, and in which she had trained the treble chorus. On 5 March, she made her debut in Handel's *Judas Maccabeus* in the Upper Assembly Rooms.[66] The significance of Herschel's first appearance in this particular work is too good to miss. First performed in 1747, *Judas Maccabeus* was conceived by Handel as a victory oratorio to celebrate the crushing of the 1745 Jacobite Rebellion, which had culminated in an exceptionally heavy defeat for the outnumbered Charles Edward Stuart and his Jacobite army at the Battle of Culloden in April 1746.[67] For Caroline Herschel, her own triumphant defeat over the expectations of others must have made *Judas Maccabeus* an especially personal debut. The soloists were advertised as Miss Mahon, Miss Herschel, Mr Parry and Mr Herschel; Caroline was therefore second principal to Miss Mahon. Herschel also listed the singers in the same order in her *First Autobiography*.[68] Within five years of teaching herself to sing, Herschel was second soloist only to a woman described as 'the celebrated Miss Mahon of Oxford'. She was almost certainly Elizabeth Mahon, one of a large and musical English family of Irish origin, and the eldest of five sisters.

Miss Mahon was also an experienced professional who had already sung several times at Bath.[69] To have come so far in such a short space of time was an impressive achievement for Caroline Herschel, who, only half a decade ago, was dreaming of escaping domestic drudgery through the improvement of her marketable accomplishments. Now, she was on the stage of one of the most popular venues in one of the most popular towns in Britain.

And she was a success. So much so, indeed, that the following year she became premier soloist. The benefit concert for William held on 15 April 1778 in the Upper Assembly Rooms, with music taken from Handel's *Messiah*, listed Caroline Herschel as first principal; Miss Cantelo as second. Miss Cantelo, who later became Mrs Harrison, was an eighteen-year-old rising star and had an impressively professional pedigree. She was an 'articled pupil' of J.C. Bach and his wife, the famous soprano Cecilia Grassi, and would later receive tuition from Rauzzini and sing in Haydn's London concerts of the 1790s.[70] Although it could be argued that William deliberately placed his sister ahead of Miss Cantelo solely for reasons of nepotism, Caroline must have been good enough to take on this challenge, especially as the concert was a benefit, whereby the profits would go to William himself. Why would he have risked losing face or finance by allowing his sister, if she was not competent enough, to take the position of first soloist?[71] The need to impress the Bath audiences would have substantially increased for William Herschel, whose succession of his old rival and despised nemesis Thomas Linley as director of the Bath orchestra in the summer of 1776 was viewed as a wholly unsuccessful move.[72] Caroline Herschel must have made quite an impact with her performance in this most popular of Handel's works.[73] As she wrote of her reception in an understated manner in the *First Autobiography*:

> I was in 1778 first Singer in the Concerts and I suppose I must have acquited myself tollerably well in the principal Songs and Recitatives of the Messia; for before I left the Rooms I was offered an engagement for the Music meeting at Birmingham. But as I never intended to sing any where, but where my Brother was the Conductor I declined the offer.[74]

Why Caroline Herschel turned this invitation down has perplexed historians of the Herschel family for over a century. With what seems to be an uncharacteristic reaction from the woman who desired to pursue her ambitions and make herself independent, it becomes hard not to agree with the saccharine conclusion of early twentieth-century assessments of Herschel's career which suggest that she gave no thought to the sacrifice she was making. Herschel was 'the pivot round which her brother's life turned, as a musician, as optician, as "explorer of the heavens".[75] Or, more recently, there is the succinct statement that: 'The chance had been offered, and spurned, and would never return'; 'her devotion to William took precedence over her desire for financial independence'.[76] Birmingham was developing as a centre of music, and its increasing industrial wealth had begun to attract singers of quality, so why did Herschel refuse the invitation? Perhaps she required the confidence which only her brother's musical direction could give her?[77]

For a woman who had desired for so long to make herself somebody in the eyes of the world, Caroline Herschel must have been delighted with the accolades she received. The prospect of singing in

Birmingham should have been an enormous boost to her hopes of independence. Several reasons can be mooted for the rejection of the offer, but none seem satisfactory. Herschel's English was still a little shaky, so maybe she was afraid of being launched in a strange place where there was no one who could translate for her. And yet Herschel related with pride the compliments on her pronunciation which were received even from the aristocracy. Herschel's 'friend', the Marchioness of Lothian, claimed that she was able to speak her 'words like an English woman'. The possibility that she was too frightened to sing without her brother, or lacked confidence in her ability, can also easily be dismissed when we examine more closely her own assessment of her vocal achievements. In a footnote to the account of her Bath singing career, Herschel detailed an incident whereby she was quite evidently accused of arrogance about her abilities. Just after her remembrance of the Marchioness of Lothian's compliment, Herschel states:

> Writing the above brings to my recollection the
> answer I gave once to a Lady, when reproving me

for being my own Trumpeter, by saying how can I help it? I cannot afford to keep one![78]

Caroline Herschel has become known as an exceptionally modest, self-effacing satellite, who spun solely around her brother, 'singing when she was told to sing, copying when she was told to copy'.[79] And yet her own recollections and the response of her contemporaries prove that this is a far from accurate representation of her own assessment of her talents. Herschel knew how much hard work she had put into perfecting her voice – why should she not blow her own trumpet? If she did not do so, no one else would. And who, after all, felt more acutely the necessity of being noticed?

Why Caroline Herschel turned down such a promising opportunity may never be known, but what we can know for sure is that she did value her own musical abilities. The fact that the 1779 season saw her fall to second soloist behind Miss Cantelo probably had nothing to do with the decline in her career. The replacement of Herschel with Cantelo was more than likely to do with the increasing powers of the highly-gifted younger woman, or the time-consuming involvement of Caroline Herschel in her brother's scientific activities.[80] The importance of astronomy in the lives of the Herschels, which had grown dramatically from their father's introductions to the mechanics of the heavens in Hanover to the construction of telescopes in Bath, was about to reach stellar proportions. For Caroline Herschel, another change in career and another attempt to gain her much-cherished independence were fast approaching.

Chapter III

From stage ornament to
celebrated female astronomer

After the decision not to pursue the Birmingham offer, Caroline Herschel still continued to take an important part in the oratorios and concerts of the season in Bristol as well as Bath. In the winter of 1779, she sang 'Italian Songs and assisted in some Glees which came then much in fashion'. William Herschel seemed to find it difficult to keep his mind on music at this time, however, and although he still composed and conducted, even his teaching reflected his now more than leisured scientific tinkering, encompassing 'lessons in Astronomy' as well as the more usual musical accomplishments.[1] The unsettled nature of the Herschels was apparent from the fact that they moved their household three times between the autumn of 1779 and the spring of 1781. From 19 New King Street in September 1777 to 27 Rivers Street in December of the same year, the Herschel siblings finally returned to the house in New King Street in March 1781.[2] On the 13th of that month, William Herschel discovered a new planet, which would later become known as Uranus, and the lives of both William and his sister would, once again, alter dramatically. With this final change of career, Caroline Herschel would achieve her ambition and earn an independent income. She would also find international fame for her own scientific discoveries.

In the late 1770s, William found his telescopes attracting renowned visitors such as Samuel Lysons, future Vice-President of the Royal Society, and Nevil Maskelyne, Astronomer Royal and later good friend and ally of both the Herschel siblings. Amusingly, William Herschel pronounced Maskelyne to be 'a devil of a fellow' after 'having several hours spiritted conversation with him', unaware of his exalted position.[3] Caroline was still busy in the world of music. The winter engagements of 1780 revealed a particularly frenetic season:

> I was obliged to rehearse the treble of the Chorus singers for the Oratorios which were to be performed both at Bath and in Bristol 16 or 18

Woman a[n]d boys were with me on those nights my Brother was at the Harpsichord in the Theater. The solo performers he engaged from London, and I only lead the treble as I did the season before; for the interruption in my practise of the preceeding months, besides accumulation of copying music &c left me no time to take care of myself or to stand upon nicity's.[4]

That Herschel no longer sang as a principal soloist is not clear evidence that her career as a singer was in decline.[5] The bringing to Bath of metropolitan singers implied that William Herschel was keen to regain his musical reputation through association with the best and most well known, even as he continued to devote more and more time to astronomy. He had allowed his sister to dominate the concerts and oratorios for a couple of seasons, so maybe he was afraid of being accused of precisely the nepotistic follies he despised in Linley. As William Herschel was aware, audience demand would be for the new and fashionable; he could not retain the same singers year after year if he was to consolidate his directorial skills. Although Caroline Herschel did not appear as a soloist, she kept an important place within the chorus as leader and director of the trebles, which was evidently a position of some responsibility. But what is apparent from this extract from *The First Autobiography* is Herschel's disgruntlement with the situation. The endless number of roles she was encouraged to fulfil did not permit Caroline Herschel to practise and perfect her own voice, nor did it allow her to continue to cultivate the social polish she had been praised for a few years before. The ornament she once was, Herschel implied, was becoming tarnished due to recurring neglect.

The possibility that William Herschel felt some regret at the way the career he had instigated for his sister was turning out can perhaps be perceived in his actions after the family moved to Rivers Street in December 1779. William took the lease of the whole house, but let the lower part of number 27 to a prospective millinery business, 'in which', Caroline noted, she 'was to have a share'. Herschel was not to 'give up more of [her] time than was necessary merely for keeping the books and see[ing] that nothing went wrong'.[6] Her brother invested a 'sum then at his disposal' to be 'laid out to advantage in the stock of the trade'.[7] For a single woman with a little capital, millinery could provide an attractive opportunity for investment.[8] William Herschel clearly thought so. Millinery was seen as a predominantly female trade, although by the end of the eighteenth century men were encroaching upon the profession.[9] Bath, with its fashionable reputation and seasonal influx of visitors, could provide a tailored audience for millinery work. William's sister had plenty of experience in looking after the Herschel household, both in Hanover and in Bath. In the latter, she had also learnt book-keeping and had put a great deal of effort into managing the section of the choir under her command. She had also been educated in the millinery profession in Hanover almost a decade before. If anyone could ensure that 'nothing went wrong' with this investment it was Caroline Herschel.

But the Herschels, or at least William, had not banked on the geographical problems of Rivers Street. After only a few months, the business collapsed due to its poor location. Situated away from the main thoroughfares, and thus away from the main custom which congregated around the centres of Bath entertainment, the business seemed doomed from the outset. Coupled with the unfavourable site, Caroline appears to have been eternally vigilant in order to 'guard against imposition and dishonesty'.[10] Whether these difficulties arose from customers or from the proprietors, Herschel does not make apparent in her account of this adventure. Yet, little was lost from the experience, as the remaining goods were successfully sold off and the business closed. This brief episode provided Caroline Herschel with further experience of a profession in which she had formerly hoped to make a name for herself, but, yet again, this was another opportunity which did not lead anywhere. Herschel still lacked the stability she so desired, and her brother could not buy her happiness, as Alexander Herschel had so depressingly predicted.

While things were looking desolate once more for Caroline, for William a final breakthrough in his attempt to make a lasting impact upon the astronomical world came to fruition on 13 March 1781 in 19 New King Street. Caroline was absent for this momentous discovery as she was winding up the millinery business in Rivers Street. The unique size and power of the Herschelian telescope allowed William to observe even the faintest objects in remote regions of the cosmos where previous astronomers had not penetrated. Before the discovery of what would later become Uranus, Saturn had been considered, since ancient times, the outermost planet. Herschel's telescope allowed him to detect an object which he initially mistook for a comet, but which in fact turned out to be a new planet.[11] He used reflecting mirrors, rather than the glass employed in refracting telescopes. In the latter, light passes through the objective glass at the mouth of the tube, which brings it to a focus where the image can be examined through the eyepiece; but in the former, the light passes down the tube to the reflecting mirror at the base, and is reflected back to the top of the tube, where the image can again be examined through the eyepiece. A reflector also avoids chromatic aberration, as all colours are reflected equally. Whereas a large glass refractor would require corresponding thickness, resulting in a gain in the ability of the telescope to gather light but a reduction of this light due to the thickness of the glass, a mirror would neither be limited in diameter nor, thus, lose light-gathering power.[12] In the 1780s, the use of glass would be hampered additionally by problems of increasingly vast expense. From June 1784, the government of William Pitt the Younger introduced the punitive and unpopular window tax. This was later extended to glass itself, resulting in near destruction for the optical industry.[13] The employment of mirrors by William Herschel in his telescopes offered the dual security of cheaper and more readily available material and increased visibility and cosmic penetration.

Caroline Herschel does not mention the discovery of the new planet in her autobiographies, although she offers a rare, exact dating for the day she returned to 19 New King Street to join her brother: 21 March 1781. She does, however, mention the after-effects of the discovery and her own part in helping her brother. From March 1781, 19 New King Street became a landmark for any interested in astronomy, much to Herschel's irritation at the invasion of her home at unreasonable hours. This constant domestic disturbance was something to which she would become far more used in the decades to come:

> But the interruption of Visitors which wanted to look through the telescopes at night, were very numerous, for since the discovery of the Georgium

Sidus, I believe few men of learning or consequence left Bath before they had seen and
conversed with its discoverer, and thought themselves fortunate when finding him at home on
their repeated visits.[14]

In addition to curious visitors, the Herschels were
bombarded with letters from the Royal Society and
the current President Sir Joseph Banks, who informed
William Herschel that he was to receive the Society's
Copley Medal for his discovery. This award was
founded in 1731 and named after Sir Geoffrey Copley,
who had donated £100 to the Royal Society in 1709
to pay for experiments.[15] When William visited London in the summer of 1782, he went to Greenwich,
where, 'star-gazing' with Maskelyne, the Astronomer
Royal, and the wealthy amateur Alexander Aubert,
he compared his seven-foot reflector with the best
the other two could offer. Aubert's observatory at
Loampit Hill, near Deptford, was furnished with the
best instruments that the best opticians of the day
– Short, Bird, Dollond, and Ramsden – could provide.[16] But Herschel's telescope, he claimed proudly
in a letter to Caroline

was found to be superior to any of the Royal Observatory. Double stars which they could not see with their instruments I had the pleasure to show them very plainly, and my mechanism is so much approved of that Dr Maskelyne has already ordered a model to be taken from mine and a stand made by it to his reflector.

'Among opticians and astronomers nothing new is talked of but *what they call* my great discoveries', noted William, adding in rather King Lear-like fashion that his 'trifles' are nothing compared to what he has yet to accomplish: 'Let me but get at it again! I will make such telescopes and see such things – that is, I will endeavour to do so.'[17] Despite the accolades and instant recognition, life went on for the Herschels. With increased attention being paid to his work, William Herschel had a reputation to maintain and desired to 'perfect his mirrors'. This could not be carried out alone, and so he co-opted his astronomical apprentice into helping him in his endeavours.

Unlike Alexander, Caroline proved hard-working and dedicated to what must have seemed a frustrating diversion from trying to maintain and stimulate her own career. 'In short', claimed Herschel, 'I saw nothing else and heard of nothing else':

> Alexander was always very alert and assisting when anything new was going forward, but wanted perseverance and never liked to confine himself at home for many hours together; and so it happened that my Brother was obliged to make trial of my abilities in copying for him Catalogues, Tables, &c and sometimes whole papers which were lent him for his perusal, of which among others was one of Mr Michel and a Cat. of Christian Mayer in Latin which kept me employed when my Brother was at the Telescope at night; for when I found that a hand sometimes was wanted when any particular measures were to be made with the Lamp micrometer &c and a fire to be kept in, and a dish of Coffe necessary during a long nights watching.

Appended to this list of tasks is the succinct conclusion: 'I undertook with pleasure what others might have thought a hardship.'[18] Although she was not yet apparently given the opportunity to use the telescopes in any professional manner, Herschel was slowly being introduced to the labours of another world. This time it was an astronomical one. For all the theory which she copied out for her brother, Caroline Herschel was also experiencing the demands of observation made upon scientific practitioners. Unlike Alex, she was becoming committed to more than just the exciting novelty of discovery. From making coffee to transcribing articles in languages she did not understand and had never been taught, to making measurements, Caroline Herschel had begun to learn a scientific trade.

While Alex Herschel seemed reluctant to devote too much time to the development of his brother's new interest, his sister, never afraid of hard work, embraced the change, if not wholeheartedly, then with her usual dedication. Far from solely expressing distaste at the activities, Herschel seemed to sense that they were undertaking something of extreme importance, of which she was content to play her part and perhaps achieve some recognition for her assistance. Around preparations for musical performances of 1782, William Herschel was attempting to construct a three-foot mirror to improve upon the range of his seven-foot reflector. As mirrors simply were not to be bought prepared at the size William wanted them, he was compelled to buy the disks and grind and polish them himself. But this was by no means the job of only one man: the whole family joined in with the casting of the mirror. Herschel stressed her own laborious contribution, which was, of course, she pointed out, greater than that of Alex:

> The mirror was to be cast in a mould of lo[a]m prepared from horse-dung of which an immense quantity was to be pounded in a morter and sifted through a fine seaf; it was an endless piece of work and served me for many an hours exersise and Alex frequent[l]y took his turn at it, for we were all eager to do something towards the great undertaking, even S[i]r W[illia]m would sometimes take the pestel from me when he found me in the workroom where he expected to find his friend; in whose concerns he took so much interest.[19]

Caroline implied, by mentioning the intervention of the Bath physician and member of the Royal Society Sir William Watson, that her own efforts were perhaps greater than those of William Herschel, whose musical commitments kept him from the workshop. Even though both attempts at casting the mirror failed, Herschel was more than keen to stress her efforts in backbreaking, tedious work more usually undertaken by workmen. She too, her words implied, had a large part to play in the continuing brilliance of William Herschel's reputation.

The 1782 season was the final time that brother and sister would appear on stage, and it would thus, although unknown to them at the time, bear witness to the final severance of the Herschel siblings from the musical profession. William Herschel, distracted by his astronomical studies, conducted a poor performance of the *Messiah* in Bristol which was lambasted by the press, who pandered to a readership intrigued and fascinated by gossip about musicians.[20] At one final event, in St Margaret's Chapel, Bath, on Whit Sunday 1782, both brother and sister gave their final performance.[21] It was perhaps with relief that William Herschel played his last note that season. He had achieved enormous popularity during nearly twenty years' residence in the city and had managed to dismiss his earlier fears of treating music as a commodity by becoming a skilled tradesman. His teaching practice rivalled that of Charles Burney's in London. Burney was nothing if not adept at puffing his own abilities, and Herschel must have been similarly socially adroit.[22] His circle of pupils became so extensive and fashionable that Caroline Herschel's supporter, the Marchioness of Lothian, could organise twenty private concerts in which, with a professional quartet, Herschel's protégés could display their talent.[23] By 1771, indeed, William Herschel was earning around £400 a year.[24] Although his earlier compositions showed some individuality, his later productions lacked inspiration. They leant towards the superficial and conventional and thus betrayed evidence that Herschel was content simply to keep his audience happy, as well as to devote his own time to his increasingly astronomical passions.

His sister, however, did not yet feel quite the same about the need to pursue astronomy. William's summons to the court of George III and Queen Charlotte led to a stay of almost three months between May and July 1782 and the final offer of a position as the royal astronomer at Windsor with a salary of half what William had made as a musician: £200 per annum. William's dedication to astronomy must have been great to take such a drastic reduction in salary. The pursuit of science in Britain at this time,

and indeed for another half a century to come, was by
no means a lucrative option. A scientific 'profession'
simply did not even begin to exist until the second
half of the nineteenth century. Indeed, discoveries
were often made by wealthy amateurs or those, like
William Herschel, whose success was achieved in the
time they could spare from another, regularly paid
employment. An independent income was usually
necessary for any attempt at scientific endeavour.[25]
Even those who earned their income from university teaching were obliged to supplement their dire
wages in order to make ends meet. In scientific terms,
William Herschel's salary was therefore generous.
Caroline Herschel did not think so, and the meanness of her brother's wages rankled with her into
extreme old age. As she was the one to whom the
responsibility of housekeeping would devolve, she
was astounded at the comparative poverty the new
position would bring the family. In a letter of April
1827, Herschel informed her nephew, William's son
John, that she was never 'satisfyed with the support
your Father received towards his undertakings'.[26] Nor
was Herschel the only dissatisfied supporter of her
brother. Sir William Watson expressed his displeasure

that the monarch had obtained William Herschel's services at 'so cheep' a rate. But, for William Herschel himself, a return to music would be 'an intolerable waste of time'.[27]

The prospect of a lower income for the Herschel family was not the sole reason for Caroline's worries about the move to a new house and a new area. As everything happened so quickly – they had left Bath and found a house in Datchet near Windsor by August 1782, barely a month after William's appointment – Herschel was unable to realise fully what she had lost in leaving Bath until they had actually moved. When Alexander returned to the city for the start of the season in October, the implications of the change in circumstances finally hit Caroline – hard:

> The beginning of October Alexander was obliged to return to Bath, the separation was truly painful to us all, and I was particularly affected by it; for till now I had not time to consider the consequence of giving up the prospect of making myself independent by becoming (with a little more uninterrupted application) a useful member to the musical profession. But besides that my

Brother would have been very much at a loss for my assistance, I had not spirit [*corrected to*: impudence] enough for throwing myself on the public after losing his protection.[28]

The move to Datchet, or so Herschel thought, effectively destroyed yet again her prospects of financial independence. It is noticeable that she did indeed consider herself talented enough to continue, and only required more time to rehearse her voice. Through constant interruption of one thing or another, she had been unable to practise in the way she had expected on leaving Hanover ten years before. Now Herschel was faced with the enforced ending of her career. While she felt that her brother needed her assistance, especially without Alexander around, it is clear that this was not the only consideration which prevented Herschel from returning to Bath alone. Obviously she was deeply grateful to William for the opportunities he had encouraged her to take by leaving Hanover, but there is something else in this statement. The fact that the word 'spirit' was crossed out may give some indication as to why Herschel believed she could not return. It would have been 'impudent' to leave William, but Herschel was distinctly unable to describe herself as lacking spirit. Ultimate success in the music profession required the suppression of precisely the kind of 'spirit' that Caroline Herschel possessed. For the eighteenth-century musician, professional life was precarious. The necessity of piecing together an income from several sources often led to mercenary behaviour, requiring knowledge of the musical market more common to trade than art. Succumbing to financial temptations, musicians could take on more work than could possibly be performed or send deputies to less rewarding concerts or gatherings. With the unprecedented opportunity to cross barriers of wealth and class, the musician also needed to adjust skilfully and with considerable tact to social situations.[29] While this ability to flatter and fawn upon one's social superiors came easily to William Herschel, as it did to other musicians, his sister presented repeated evidence in her autobiographies of her inability to conform to what she labelled the necessity to 'stand upon nicity's'.[30] Better to stay where she was than to have someone reminding her, as William frequently did, of her lack of toadying in displays of *faux* public decorum.

Deciding to remain where she was, Herschel was confronted with how to survive on £200 a year. Although the Herschels had been born into humble stock where this would have been a great deal of money, they had been used to double this amount in Bath. Herschel was astonished with the inflated expense of living nearer to the metropolis. The slightest thing sent her into apoplexies:

[T]he first time I went to markets I was astonished at the dearness of every article and saw at once that my Brothers Scheeme of living cheep in the country (as jokingly he said) on Eggs and Bacon would come to nothing; for at Bath I had the week before bought from 16 to 20 Eggs for 6^d, here I could not get no more than five for 4^d Butchers meat was 2 ½^d and 3^d per pound dearer; and the only Butcher at Datched would besides not give honest weight and we were obliged to deal all the time we lived there at Windsor. Coals were more than double the price. O dear! thought I what shall we do with 200 a year! After Rent and Taxes are deducted; the lat[t]er were in consequence of the overcharged rent (and there being upwards of 30 Windows on the premises) enormeous; and notwithstanding all the labour and expense bestowed upon it, it remained an incomfortable habitation.[31]

Mrs Papendiek, a fellow German and Assistant Keeper of the Wardrobe and Reader to Queen Charlotte, offered some corroboration for Herschel's estimates in her own account of this period at Windsor:

While I am speaking of expenditure it may be interesting if I mention the prices of provisions and other necessities of life in those times. Meat, taking one kind with another, was fivepence a pound; a fowl, ninepence to a shilling; a quartern loaf, fourpence; sugar, fourpence a pound; other groceries about the same as now, except tea, which was very much more expensive – 6d a pound and upwards.

I should say that as a rule ordinary everyday things were cheaper, and luxuries decidedly dearer; but people were content without them, and were not despised for living economically.[32"]

Both narratives reveal how women were compelled

to spend income carefully, and simultaneously how much responsibility fell upon their shoulders for running the household smoothly. As the conduct-book writer Lady Sarah Pennington had stated in *An Unfortunate Mother's Advice to her Daughters* (1761), proficiency in accounts was absolutely vital for woman's control over her household in the eighteenth century; one domain where she could and should dominate.[33] Teaching women even the most basic elements of mathematics, as more people began to realise during this period, was essential for the maintenance of domestic harmony.

But William Herschel intended to teach his sister more than simply basic arithmetic. From almost the very moment they arrived at Datchet, he revealed that Caroline's position in the Herschel household would involve more than the female duties she had learned to perfect over the years. Alexander's recent marriage, his return to Bath and his essential reluctance to help to any extent in the workshop had effectively ruled out his long-term co-operation in astronomical activities at Datchet. Caroline, however, was a different matter. She had proved herself willing and able to take on even the menial tasks involved in the production of telescopes and the taking of measurements during observations. She was certainly a quick learner – her successful singing career had more than illustrated that. And she had shown interest in astronomy from the outset of their life together in Bath. Indeed, as we saw in the first chapter, Caroline Herschel had been introduced to the wonders of the celestial world from a very early age by her father Isaac. Almost as soon as the siblings arrived in Bath, Herschel was involved in conversations about astronomy with her brother. At the time, they talked, 'by way of relaxation', about 'the fine constellations with whom I had made acquaintance during the fine nights we spent on der Postwagen traweling through Holland'.[34] It even seems possible that William Herschel had tried to entice his sister into observing with him in Bath. This encouragement occurred in 1779, the year that Caroline slipped to the position of second soloist in concerts and oratorios and the year in which William Herschel bought a share in the millinery business. It appears, therefore, that the brother sought to encourage the despondent sister in a potentially new career from as early as April 1779, when he stated himself in notes concerning

'Experiments on the construction of specula' that he 'repolished both speculums of my sister Carolina's Gregorian telescope this morning'. An independent observer also corroborated this evidence. Dining with the Herschels one evening, John Marsh noted that William had given him the impression that 'His sister and his brother ... were as fond of astronomy as himself and all used to sit up, star-gazing, in the coldest frosty nights'.[35] Alex's reluctance to put himself out in any way for scientific endeavours requiring persistence or hardiness make his participation unlikely, but there is no reason why Caroline Herschel should not already have been shown how to observe by her brother, especially in the light of his apparent guilt over the trajectory of his sister's singing career.

The speed with which William co-opted his sister into his astronomy does indeed suggest that Caroline had used a telescope before, even if it was only as a leisure pursuit in the very few moments she had to spare from her musical and domestic duties. Within a few weeks of their move to Datchet in August 1782, Caroline Herschel was observing the skies. And this time, without her singing career to distract her, Herschel was to devote herself to astronomical observation. Already representing herself as an apprentice in the workshop, Herschel was now to receive more 'training', this time as 'an assistant Astronomer'. William Herschel had clearly decided that his sister would provide him with the best – and naturally unpaid – assistance he could find. The way she phrased this dramatic change in her circumstances shows how easily her brother could dictate his wishes to Herschel, who was still dependent, to her evident chagrin, upon his protection. Her thoughts, she noted depressingly, 'were anything but cheerful':

[I] found I was to be trained for an assistant Astronomer and by way of encouragement a Telescope adapted for sweeping consisting of a Tube with two glasses such are commonly used in a finder, I was to sweep for Comets, and by my Journal N° 1, I see that I began Aug 22^d, 1782 to write down and describe all remarkable appearances I saw in my Sweeps (which were horizontal). But it was not till the last two months of the same year before I felt the least encouragement for spending the starlit nights on a grass-plot covered by dew or hoar frost without a human being near enough to be within call; for I knew too little of the real heavens to be able to point out every object for finding it again without losing too much time by consulting the Atlas.[36]

By employing the passive voice, Herschel revealed immediately that she was not consulted about her new role. The bitterness of the rest of the passage showed her anger at the alteration in her lifestyle. After all, not even the nights were her own any more. But Caroline Herschel was not one to give up so readily on her new role in life. At first reluctant to spend hours isolated in cold and wet surroundings, Caroline soon warmed to this new challenge, even though her knowledge of abstract mathematical and astronomical calculation was non-existent. Responding to an 1842 letter from Margaret Herschel about how her husband John was teaching their children mathematics, the 92-year-old Herschel recorded the novel method employed by William in instructing his sister in geometry. Even though she was a 'grown woman', Herschel commented wryly, 'I sometimes fell short at dinner, if I did not guess the angle right of the piece of Pudding I was helping myself to'.[37]

But the lessons did become harder. Once again, she was determined to master her new art, and bombarded her brother with complex 'inquiries' over breakfast – as at Bath, the only time they could spend together before they 'separated each for our daily tasks'.[38] She recorded the answers to her queries and other speculations in her 'Commonplace Book', which contained examples of how to take equal altitudes, how to convert sidereal time into mean time, how to find the logarithm of a number given, how to calculate oblique plain triangles and right-angled spherical triangles, and theorems for making tables of motion.[39] While she drew her nephew's attention to his father's 'perseverance', she also makes the reader aware of her own sacrifices, her own ability to 'work through every obstacle which came in [her] way'.[40] By the end of the year and the beginning of 1783, Herschel's feelings towards her allotted career had begun to alter dramatically. Ironically, despite complaints about the harsh weather, Herschel started to enjoy herself in the coldest months of November and December. The close, if distracted and absorbed, company of her brother was also a huge boost, not only to her feelings of loneliness but also to her observational development. Alongside descriptions of her lack of scientific knowledge, Herschel revealed simultaneously her increasingly impressive command of astronomical observation. '[I] could have his assistance immediately', she wrote in her autobiography, 'when I found a Nebula or Cluster of Stars, of which I intended to give a Catalogue'.[41] From confused novice who could not tell one celestial object from another without consultation, who had passively 'found herself ' an astronomer's assistant, Herschel had in fact found fourteen nebulae by the end of 1783 and was set to record a catalogue of the findings. William Herschel had been spurred on to seek for further nebulae (Latin for 'mist' or 'clouds') when he was given a copy of Charles Messier's catalogue of 100 known nebulae.[42] Known as the 'ferret of comets' by virtue of the number he had discovered, Messier had been observing a comet in Taurus, when he found the Crab nebula (M1). This led to the decision to form a catalogue of nebulae and clusters to facilitate comet-searching. By 1784, Messier had drawn up a list of 103 objects.[43] The Herschels thus contributed significantly to the development of cosmological thought. With the improvement in the manufacture of telescopes – in which William Herschel had been a pioneer – interest grew in nebulae themselves, which, thanks to Messier's catalogue, had been more successfully distinguished from comets. William Herschel brought both himself and his sister to the forefront of astronomical innovation, in spite of their relatively recent and autodidactic acquaintance with the heavens. The latter's achievement was undeniably astounding. Barely a year after she had reluctantly begun her observations, Caroline Herschel had increased the number of known nebulae by 5 per cent. And this with a telescope little better than a child's toy.[44]

Herschel's 'Books of Observations' were filled with her early discoveries. The year 1783, like 1778 when she became premier soloist at the Bath oratorios, was to be a memorable year for Caroline Herschel. On 26 February she recorded the dramatic events of the evening. Two nebulae had been discovered previously by Messier (M46, M41), but another two found by Caroline were unknown to astronomy. The double conclusion 'Messier has it not' concealed the magnitude of Herschel's find. In a few minutes, Herschel had increased the number of known nebulae by 2 per cent.[45] Persistence had paid off; Caroline

Herschel was now a practising astronomer in her own right with scientific discoveries to her name. William must have been delighted with the progress of his sister. Yet again, she had proved how impressively she could adapt to novel circumstances and succeed in her allotted tasks. The astronomical finds in Herschel's observation book may have been dictated to her by her brother, but the discoveries were hers.

As a reward for her considerable achievements, William replaced Caroline's telescope with a more powerful instrument, which she first employed on 8 July 1783. Her first instrument was a simple tube with two glasses which pivoted about a vertical axis, making it suitable for horizontal sweeps. A handle wound in the string which would allow Herschel to raise the tube without moving her eye from the eye-piece. The new model, by contrast, was larger, a Newtonian reflecting telescope of just over two feet focal length, which usually had two mirrors, but which had been adapted by William to contain only one, thus allowing the incident light to be reflected from the speculum directly onto the eyepiece without further loss of brightness.[46] The Astronomer

Royal, Nevil Maskelyne, who had concluded in 1782 that William's telescopes were in fact more powerful than those at Greenwich, wrote favourably of a later five-foot version of Caroline Herschel's instrument in a letter to Edward Pigott of 6 December 1793:

I paid Dr & Miss Herschel a visit 7 weeks ago. She shewed me her 5 feet Newtonian telescope made for her by her brother for sweeping the heavens. It has an aperture of 9 inches, but magnifies only from 25 to 30 times, & takes in a field of 1° 49' being designed to shew objects very bright, for the better discovering any new visitor to our system, that is Comets, or any undiscovered nebulae. It is a very powerful instrument, & shews objects very well. It is mounted upon an upright axis, or spindle, and turns round by only pushing or pulling the telescope; it is moved easily in altitude by strings in the manner Newtonian telescopes have been used formerly. The height of the eye-glass is altered but little in sweeping from the horizon to the zenith. This she does and down again in 6 or 8 minutes, & then moves the telescope a little forward in azimuth, & sweeps another portion of the heavens in like manner. She will thus sweep a quarter of the heavens in one night. The Dr has given her written instructions how to proceed, and she knows all the nebulae [listed by Messier] at sight, which he esteems necessary to distinguish new Comets that may appear from them. Thus you see, wherever she sweeps in fine weather nothing can escape her.[47]

Within a few years, these telescopes would help to catapult Caroline Herschel into international fame as an astronomer.

But before she had much of a chance to use her instrument, Herschel told a typical story. As with every other time she had tried to improve her knowledge of her current occupation, something prevented her. This time it was William's need for her assistance in writing down his own observations with his latest telescope, the 'large' twenty-foot, with its eighteen-inch mirrors and additional chair, '15 or &c feet' above the ground, which allowed the observer to work without the constant fear of falling.[48] His attempt both to observe and to record the results of the observations proved time-consuming and eventually impossible. In order to carry out both tasks, William needed his sister to be a constant presence at her desk. As well as amanuensis, Herschel was employed as general dogsbody by her brother, and as a consequence, she wrote depressingly, her 'Sweeping was interrupted':

> But it could hardly be expected to meet with any Comets in that part of the heavens where I swept, for I generally chose my situation by the side of my Brother's instrument that I might be ready to run to the clock or write down memorandums. But in the begining of December I became entirely attached to the writing desk and had seldom an opportunity after that time of using my new acquired Instrument.[49]

Herschel did not seem to mind the constant requests to fetch and carry, although she lamented the time she was forced to spend from her instrument. In her 'leisure hours', Caroline returned to the workshop to continue the labours she had begun in Bath, and 'ground 7 feet and plain mirrors from ruff', even delighting proudly in the 'polishing and last finishing of a very beautiful mirror' for Sir William Watson.[50]

Comforted by William's need of her, Herschel ran to the clocks, wrote down memorandums, fetched and carried instruments, measured the ground with poles, '&c &c'. She certainly filled her role as apprentice, compelled to be on the alert constantly because 'something of the kind every moment would occur'.[51] Unsurprisingly, in such a fraught and frantic atmosphere, accidents were bound to happen, and William had already had a lucky escape when, in a high wind, with the structure of the twenty-foot still not securely in place, the whole apparatus came down, luckily neither smashing the mirror nor crushing William Herschel.[52] On 31 December 1783, it was Caroline's turn to succumb to the pressures of the system of observation. The night had begun cloudy, but around ten o'clock the sky cleared and a few stars were visible. In 'the greatest hurry all was got ready for observing'. In *The First Autobiography*, Herschel explained how the accident occurred:

My Brother at the front of the Telescope directing me to make some alteration in the lateral motion which was done by a machinery in which the point of support of the tube and mirror rested

...[,] at each end of the machine or trough was an iron hook such as butchers use for hanging their joints upon, and having to run in the dark on ground covered foot deep with melting snow, I fell on one of these hooks which entered my right leg about six inches above the knee, my brothers call make haste I could only answer by a pittiful cry I am hooked. He and the workmen were instantly with me, but they could not lift me without leaving near 2 oz. of my flesh behind.[53]

Faced with the incompetence of her brother, the workman who raised and lowered the telescope tube and his wife, Herschel was forced to administer surgery to herself, by applying 'aquabaseda and tying a kerchief about it for some days'. In spite of what must have been horrendous pain and a perpetual fear of gangrene setting in, Herschel did not consult the local doctor, who eventually 'heard' of the accident from another unnamed source and brought ointment and lint. After six weeks, the doctor returned to find the wound healing and to inform his patient that 'if a soldier had met with such a hurt he would have been entituled to 6 weeks nursing in a hospital'.[54]

Remembering this conversation, Herschel emphasised not only her own incredible stoicism, but, in the
military link, her sense that she had been wounded
while on professional duty. Her thoughts were with
her brother, who might have to do without her for a
few nights, but the next two weeks were fortunately
cloudy and William was unable to observe for any
length of time. While this episode has been frequently
employed by historians of the Herschel family to
illustrate Caroline's utter devotion to her brother,
it is only too noticeable that she is in fact extremely
glad of what, quite clearly, was much-needed rest, as
well as completely convinced of her indispensability.
Never one to stand upon those social 'nicity's', Caroline Herschel realised the extent of her importance
to her brother, who in turn represented the world of
astronomy. Her contribution, as she very well knew,
was vital to William's continued success.

William Herschel's 'chief object' in the mid 1780s
was to construct a 30- or 40-foot reflector, which
would be the largest telescope to date in the world.[55]
In order to achieve this objective the Herschels
needed money from the King, which was not easy
to obtain, nor very forthcoming. As the £200 annual

salary did not cover the expenses of living and constructing telescopes for other people, William needed more money to aid the development of what he eventually decided would become a 40-foot reflector. In the summer of 1785, the Herschel siblings moved from the 'damp' Datchet house, which was making them ill, to a smaller property, Clay Hall, in Windsor. They lived here until April 1786, when they moved to Slough, forced out by a landlady who kept raising the rent.[56] If the time spent at Clay Hall proved unsatisfactory in some ways, in others it was more successful. In September 1785, the King granted William the £2,000 he had required in order to begin construction of the 40-foot. The President of the Royal Society, Sir Joseph Banks, had, not for the first nor last time, been William's intermediary in achieving the funds for the project. While work finally began on the 40-foot at Slough, the summer of 1786 would witness further innovation, and this time it would be Caroline Herschel who would be fêted nationally and internationally for her scientific discoveries.

Having helped her brother to the detriment of her own sweeping for the past couple of years, Herschel relished the time she had to herself when he was absent. In July 1786, William was sent by the King, along with Alexander, to deliver a ten-foot telescope as a present to the Observatory of Göttingen, in his home state of Hanover.[57] At first, Caroline was lonely and sad, recording in her 'Book of Work Done' that she 'began with bustling work' in order to forget her isolation. Mostly, she recorded her domestic duties, as she cleaned, sewed and shopped, as well as dealing with the machinations of an idle gardener and showing around some visitors, as she put it, who had come to see her.[58] Clearly, others already found this self-educated female observer just as fascinating as her brother. The visitors ranged from academics and astronomers to the aristocracy and royalty to whom the King demanded she show the instruments.[59] As member of the royal household Mrs Papendiek remembered in her memoirs, the twenty-foot was particularly popular: 'Company and friends were never denied admittance to view this extraordinary piece of mechanism, nor in the evening to look at the moon and planets, either through the large telescope, or another of 10 feet, also fixed in the garden.'[60] The month passed in similarly repetitive fashion, as she worked and entertained over and over again. Alone, with only Alexander's irritatingly gossipy wife for company, Herschel felt how anomalous her position in the house truly was. She was neither 'Mistres of her Brother's house nor of her Time', and was therefore unable to give 'infitations'. Although occasionally and evidently interested in those visitors who came 'to see her', Herschel was at a loss, with her inability to stick rigidly to the 'nicity's', how to deal effectively with such 'self inviting visitors'.[61] The lack of freedom and a secure establishment rankled endlessly with Caroline Herschel.

As Herschel continued to calculate the catalogue of nebulae which she and her brother had been working on since 1783, she also found that she had, finally, some time to herself. In a letter to John Herschel's wife, written almost 60 years after this period, Caroline Herschel remembered precisely how liberated she had felt when sweeping alone, 'the many solitary (and at the same time) happy hours spent on my little roof at Slough when I was not wanted at the 20 feet'.[62] Left to her own devices, Herschel discovered her first comet on 1 August 1786. It is significant that this observation was made when William was absent, and it belies the suggestion made previously by historians of the Herschels that Caroline had little or no interest in astronomy. She did not have to sweep when William was absent, and the time she was able to take from her usual duties as amanuensis she could have spent in any fashion she wished. Instead she chose to continue, in a professional manner, and profited from her observational perseverance. For the next few days, the house at Slough was bombarded with letters of congratulation and important visitors, after Herschel had announced her discovery to the world. In her letters to notable astronomers, Herschel emphasised not only her lack of observational practice but the time she had spent aiding her brother to her own disadvantage. The minute William had left for Germany, his sister moved to boost her own talents – and in doing so, achieved a find of which even she can scarcely have dreamed. On 2 August, Herschel wrote to Alexander Aubert, the wealthy amateur who had found William's telescopes more powerful than any of his own: 'I hope, sir, you will excuse the trouble I give you with my wag [vague] description, which is owing to my being a bad (or what is better) no observer at all. For these last 3yrs I have not had an opportunity to look as many hours in the telescope.'[63] It is notice-able that Caroline does not blame herself for her lack of skill, but rather the fact that her time has been taken up writing down observations for her brother – her lifelong inability to perfect her talents. To Dr Charles Blagden, Secretary to the Royal Society, Herschel was even more explicit: 'The employment of writing down the observations when my brother uses the 30-ft reflector does not often allow me time to look at the heavens, but as he is now on a visit to Germany, I have taken the opportunity to sweep in the neighbourhood of the sun in search of comets.'[64] The supposedly downtrodden Caroline Herschel who 'had no ambition', whose 'distinction came to her unsought and incidentally',[65] evidently and deliberately sought to employ the time she was granted when not at the beck and call of her brother to benefit herself and achieve something from her sweeping.

The scientific community, such as it was at the time, was delighted with Herschel's find. Between 3 and 16 August, when William and Alexander returned, Herschel wrote down a list of the important visitors who came to see her and her find in her 'Book of Work Done'. Finally, she must have felt that she deserved this distinction; even though she was not mistress of the house, she could claim the comet as her own. Herschel was visited by an impressive list of notable scientists, including the President, Sir Joseph Banks, and Secretary, Charles Blagden, of the Royal Society, the Neapolitan physicist and Fellow of the Royal Society, Tiberius Cavallo, and Christian Gottlieb Kratzenstein, professor of physics and medicine at the University of Copenhagen.[66]

Before the eighteenth century, comets had been found by accident. As the philosopher and political economist Adam Smith wrote in *The Principles which Lead and Direct Philosophical Enquiries; Illustrated by the History of Astronomy*, published posthumously in 1795, comets had been 'least attended to by Astronomers' before this period: 'The rarity and inconstancy of their appearance, seemed to separate them entirely from the constant, regular, and uniform objects in the Heavens, and to make them resemble more the inconstant, transitory, and accidental phaenomena of those regions that are in the neighbourhood of the Earth.'[67] Messier's 'ferreting', however, had instigated a more systematic search for the objects by astronomers.[68] Messier had so far led the cometary field, and this British victory, as many of those who came to Slough must have realised, was considered exceptionally important for the nation. Caroline Herschel's achievement, wrote Nevil Maskelyne to William in 1786, was a triumph for British astronomy: 'I hope that we shall by our united endeavours get this branch of astronomical business from the French, by seeing comets sooner and observing them later.'[69] Alexander Aubert, replying to Caroline's letter informing him of her discovery, both praised William and noted his sister's professionalism and achievement of lasting fame:

> I wish you joy, most sincerely on the discovery. I am more pleased than you can well conceive that you have made it, and I think I see your *wonderfully clever* and *wonderfully amiable* brother, upon the news of it, shed a tear of joy. You have immortalized your name, and you deserve such a reward from the Being who has ordered all these things to move as we find them, for your assiduity in the business of astronomy, and for your love for so celebrated and so deserving a brother.[70]

From ancient times, comets had been viewed as

portentous: 'celestial hieroglyphics of God's message and atmospheric threats of his future actions'. Eighteenth-century cometary theory suggested that comets were in fact natural agents manipulated by God to achieve his designs.[71] For Aubert, God's design had been to reward Caroline Herschel for her astronomical 'assiduity', as well as her devotion to and generous participation in her brother's cause.

While amateur and professional astronomers rejoiced, the general public expressed their intense interest in Caroline Herschel's discovery. The timing was particularly fortuitous. Female ability in the arts and sciences had been highly praised earlier in the year by the *European Magazine*, and Herschel's comet find contributed to the air of cultural and national pride in the achievement of women, at its peak in the 1770s and 1780s.[72] An article on the poet Anna Laetitia Barbauld in the *European Magazine* proudly claimed the right to recognition of many more distinguished women:

On a retrospective view of those names which are entitled to literary honours, and which will hereafter redound to the reputation of the country, are to be found those of many females who have successfully explored the recesses of science, have enlarged the bounds of human knowledge, and added to the innocent and improving amusements of life.[73]

Another member of the royal household, the novelist Frances Burney, then lady-in-waiting to Queen Charlotte, went to Slough to observe Herschel's comet, because, as she stressed, the find was momentous. It may have been 'very small, and had nothing grand or striking in its appearance', stated Burney, but 'it is the first lady's comet, and I was very desirous to see it'.[74] German novelist Sophie von La Roche felt similarly proud when she met Herschel a few months after her discovery. The Herschels were, indeed, high on her list of things to see while in London in 1786. Even before she met them, La Roche had delighted in Caroline Herschel's success, mentioning how involved she had been in the maintenance of her brother's reputation, as well the impressive nature of her own achievement:

[T]his great man's sister, who accompanies him on his path to glorious immortality, not only …

help[s] him in his calculations, but in his absence recently discovered a new comet, and enjoys a claim to her brother's great reputation in the matter of the telescope discovered and perfected by him.[75]

Clearly, the fact that Herschel had discovered the comet while alone and thus without the fraternal prop was common knowledge in Britain that year, as was the laborious work she had carried out in the service of her brother. As befitting her status as sentimental novelist, La Roche almost swooned when, at her request, Caroline Herschel picked some daisies which grew at the base of the twenty-foot, for they were plucked by 'such a hand'.[76] The eighteenth century witnessed a Europe-wide craze for relics of famous contemporaries, and La Roche revealed she was no exception to this desire for collecting talismans. In similar fashion to the novelist and educational writer Maria Edgeworth, who in 1820 experienced a thrill of inspiration by visiting the house of Germaine de Staël and sitting in the chair where genius had sat,[77] Sophie von La Roche was overcome when she sat at Caroline Herschel's telescope, and felt 'real sympathy' for 'this noble creature': '"How often an important personage is replaced and nothing of importance done; from the thrones of the mighty down to the good craftsman's latest habitat. It is not the place that counts then, but a soul replete with knowledge."'[78] For those who came to marvel at the astronomical achievements of Caroline Herschel, the experience was truly inspirational. As von La Roche realised, the importance of the discovery lay in the observer herself, rather than in her location or her instrument.

The importance of Caroline Herschel was further recognised in a way she had dreamed of since her childhood years in Hanover. In August 1787, the King granted William another £2,000 for the continuing construction of the 40-foot reflector. Granting what was insisted would be the final bequest to William, George III added a provision that Caroline Herschel be paid a salary for her assistance to her brother. Herschel thus became the first woman in the history of science to be paid for her services to astronomy. This move was welcomed warmly, if a little sarcastically, by the woman who for so long had desired her financial independence:

And a salary of 50 pounds per year [Footnote: Exactly the sum I saved my Brother at Bath in writing Music by a clean fire side.] was settled on me as an assistant to my Brother, and in October I received £12 10s being the first quarterly payment of my Salary. And the first money I ever in all my life thought myself to be at liberty to spend to my own liking. A great uneasiness was by this means removed from my mind.[79]

Caroline Herschel had achieved her dearest ambition, but the regret that this had not come sooner was all too apparent in this short statement. She also hinted that she could easily have earned this money in Bath copying music in far more congenial circumstances. This was an attack not only upon her brother, who had saved money by employing his sister as an unpaid assistant, but upon the King for the meanness of the salary, given the time and effort involved. Caroline clearly felt that her labours demanded more compensation. However, in comparison to other working women of the period, Herschel earned a very generous amount. At one point she herself had desired to attain the position of governess, but knew she needed precisely what she did not have, namely the French language, in order to succeed. But a governess earned around £20 to £30 a year. In 1787, at the very top of the scale, Mary Wollstonecraft, when she worked for Lord Kingsborough in Ireland, earned £40 per annum.[80] But wages were much less for servants – positions in which Caroline Herschel had slaved unpaid since her earliest years. By the end of the eighteenth century, female domestics were paid an annual income of between £10 and £20 if they were upper servants, such as ladies' maids, companions and housekeepers, between £7 and £15 as cooks, but frequently less than £10 as housemaids, scullery maids and maids-of-all-work.[81] In many ways, Herschel's escape from the world of 'female professions' had been a lucky one. As we have seen, before the increasing professionalisation of the sciences from the 1830s onwards, the pursuit of natural philosophy was rarely rewarded as a civil position. Caroline Herschel thus achieved the double distinction of being paid for her services to astronomy and being the first woman in Britain to earn money through scientific discovery.

Indeed, it had been Herschel's suggestion in the

first place – hardly surprising, given her long obsession with the value of independence – that she be paid wages for her work. In a letter written when she was 74, Caroline reflected on her decision to request that she be paid rather than allow her brother to 'make her independent' – and thus in fact only make her more dependent on her brother's good will. Herschel wanted to labour for her wages: '[I] desired [William] to ask the King for a small salary to enable me to continue his assistant. £50 were granted to me, with which I was resolved to live without the assistance of my Brother.'[82] William's letter to the King revealed the expediency in paying his sister for her assistance to his cause:

> You know Sir, that observations with this great instrument cannot be made without 4 persons: the Astronomer, the assistant, and 2 workmen for the motions. Now, my good, industrious sister has hitherto supplied the place of assistant, and intends to continue to do that work. She does it indeed so much better, to my liking, than any other person I could have, that I should be very sorry ever to lose her from that office. Perhaps our

gracious Queen, by way of encouraging a female astronomer, might be enduced to allow her a small annual bounty, such as 50 or 60 pounds, which would make her easy for life, so that, if anything should happen to me she would not have the anxiety upon her mind of being left unprovided for. She has often formed a wish but never had the resolution of causing an application to be made to her Majesty for this purpose; nor could I have been prevailed upon to mention it now, were it not for her evident use in the observations that are to be made with the 40 feet reflector, and the unavoidable encrease of the annual expences which, if my Sister were to decline that office would probably amount to nearly one hundred pounds more for an assistant.[83]

The anxiety of Caroline Herschel's voice is apparent throughout this section of the letter; the desperate desire for independence and financial security have been reproduced for the sister by the brother almost verbatim. It is also noticeable, first of all, that William Herschel requested a sum ranging from £50 to £60 and that the King granted the lower amount; no wonder that Herschel was not entirely satisfied with her reward. Secondly, Caroline's participation in astronomical activities would cost far less than the hiring of another, presumably male, assistant. And, finally, while William Herschel stressed her status as underling, he did label her a 'female astronomer', which reiterated her standing in the astronomical world, even if it did not mention her own independent discoveries.

By the autumn of 1787, Caroline Herschel had achieved considerable distinction, in the world at large as well as in the smaller scientific community. The little girl who had dreamed of escaping her mother's oppression and tedious household tasks had achieved her ambitions at the age of 37, and now earned her own money as an assistant to her brother. While she was not paid directly for her own astronomical discoveries, her reputation as a 'female astronomer' and the publicity of her finds ensured that others knew of and applauded her independent achievements. But Herschel's newly independent status was soon to receive both a boost and a blow. She recorded her last entry in *The First Autobiography* for 8 May 1788. This was the date of William Herschel's wedding to Mary Pitt, and the effective dismissal of Caroline from her job as the Herschel family accountant and housekeeper. From this momentous date, after serving her brother for fifteen years in the fields of music and astronomy, she was truly alone.

Chapter IV

Distinguished at Last

William Herschel's marriage affected his sister so dramatically that she stopped recording her memories. The final sentence in her *First Autobiography* reads:

> And the eighth of that month being fixed for my Brothers marriage; it may easily be supposed that I must have been fully employed (besides minding the heavens) to prepare every thing as well as I could, against the time I was to give up the place of a Houskeeper which was the 8th May. 1788.[1]

That Caroline Herschel – who rarely specified exact dates unless they were especially momentous – mentioned the date twice in one sentence, indicates the dramatic nature of the event of William's marriage. But it is noticeable that Herschel referred only to her loss of domestic status. Now she received a salary for her astronomical work, she had, finally, a long-desired professional role to fulfil. On 21 December 1788,

Caroline Herschel discovered her second comet, although that comet 'ferret' Charles Messier had anticipated her.

The 1790s, a cataclysmic decade in European history, would be similarly significant for the 'female astronomer'. Further cometary finds would occur on 7 January and 17 April 1790 (the latter discovered, tellingly, while William and Mary Herschel were on 'a little tour into Yorkshire'),[2] 15 December 1791, 7 October 1793, 7 November 1795 and 14 August 1797 (with William again absent). This final comet encouraged a flurry of activity in the normally sedate Caroline, who, with 'so little faith in the expedition of messages of all descriptions', sped to London after one hour's sleep to announce the find herself; although, unfortunately, she claimed in a letter to Sir Joseph Banks,

> I undertook the task with only the preparation of one hour's sleep, and having in the course of five years never rode above two miles at a time, the twenty to London, and the idea of six or seven more to Greenwich in reserve totally unfitted me for any action.[3]

An awareness of the unfeminine lack of propriety in such an action sent Herschel scuttling home, but the initial impulse revealed her zeal for her own discoveries. As Sir Harry Englefield wrote on Christmas Day 1788 to William, after Caroline's find four days earlier, she had begun to set the world of astronomy alight and deserved all the accolades with which she was showered: 'She will soon be the great comet finder, and bear away the prize from Messier and Mechain.'[4]

Herschel's lifelong admirer, the French astronomer Lalande, paid her an intense compliment when he named his daughter Caroline. This notice was especially pleasing for Herschel, as Lalande's son was called Isaac, after Newton. This was an incentive to continue succeeding in astronomy, and a welcome boost to her activities. As she put it: 'I look upon this mark of your esteem however, as an incitement for spending what life and health may yet fall to my share, in the service of this noble Science.'[5]

Increasingly, her achievements were even being lauded in the popular press of the day, and it is likely that she appeared in a satirical print of February 1790 entitled 'The Female Philosopher smelling out the Comet'. This suggests her cultural centrality at the time and the uniqueness of her position as female astronomer, as well as the contemporary willingness both to praise and to mock women's more 'masculine' feats. *The Ladies Diary*, a unique mathematical magazine for women, noted Herschel's prowess in 1791, when it presented some 'Remarks on Comets' and discussed the discovery of her first two comets of 1786 and 1788.[6] *The New Annual Register for 1793* and *The Scientific Magazine, and Freemason's Repository for 1798* announced Herschel's finds of 1793 and 1797 respectively.[7] Caroline Herschel's name and discoveries also appeared in some unlikely sources such as *An Historical Miscellany of the Curiosities and Rarities in Nature and Art* (1794–1800) and *John Payne's Geographical Extracts* (1796). The former mentioned four of Herschel's comets, while the latter noted her discovery and sole 'scientific account' of 1793.[8] And although she appeared in *British Public Characters of 1798* under the entry for her brother, and was described as 'materially assist[ing]' William, this guide to famous contemporaries claimed that Caroline Herschel had 'distinguished herself greatly by her application to this sublime study, and has communicated to the Royal Society some very ingenious reports of observations made by her upon the starry orbs'.[9] While her importance in the domestic life of her brother diminished, her place in the history of astronomy and in late eighteenth-century culture was becoming, correspondingly, ever more secure. In a letter to the Astronomer Royal, Nevil Maskelyne, Herschel revealed just how much the pursuit of comets meant to her, 'because I have no other means (than pointing out one of those objects) of proving myself in the land of the living'.[10] Thus, recognition in astronomy effectively confirmed her existence. Astronomical sweeping would now replace the household variety in the life of Caroline Herschel.

The Herschel partnership was certainly not severed by William's marriage, although it did have several problems to contend with directly after the event. Mrs Papendiek, a close friend of Mary Pitt, now Mary Herschel, recorded in her memoirs the ups and downs created by William's marriage. For one, there would now be two establishments: Upton, where Mr and Mrs Herschel would live; and Slough, where Caroline Herschel would remain. It seems that there was mutual animosity between Herschel and her new sister-in-law, and a considerable lack of understanding on the part of the wife for the astronomical undertakings of the husband and his sister. Mrs Papendiek noted that:

> They were to live at Upton, and Miss Herschel at Slough, which was to remain the house of business. All at once it struck Mrs Pitt that the Dr would be principally at the latter place, and that Miss Herschel would be mistress of the concern, and considering the matter in all its bearings, she determined upon giving it up. All at once Dr Herschel expressed his disappointment, but said that his pursuit he would not relinquish; that he must have a constant assistant, and that he had trained his sister to be a most efficient one.[11]

William Herschel won the argument. There were to be two maidservants in each establishment, and they would share a footman, who would go backwards and forwards between Slough and Upton. 'Miss Herschel', Mrs Papendiek stated, 'was to have apartments over the workshops'.[12] Clearly, the thought of her sister-in-law having responsibility as mistress of the house of 'business', where her new husband spent most of his time, irked Mary Pitt considerably.

Although there was evidently no love lost between the two women in their fight for William Herschel's affections, later they became very good friends and correspondents, and Herschel much lamented her sister-in-law's death in 1832.

But what seems to have concerned Caroline Herschel most about the upheaval in her situation was, unsurprisingly, the sheer disruption caused to her working life. While the recording of her memories in later life ceased in the year 1788, there are other, contemporary accounts and memoranda kept by Herschel in the form of a day book composed between 1797 and 1821 which give an insight into the problems of the disintegration of the Herschel siblings' domestic arrangements. In a memorandum of 31 December 1798, Herschel wrote: 'Uncommonly harassed in consequence of the loss of time necessary for going backward and forward, and not having immediate access to each book or paper at the moment when wanted.'[13] Serious frustration resulted for this meticulously-minded woman. On 28 March 1800, Herschel cried:

The MSS and astronomical books in general were removed out of the observatory above stairs and lodge in my brother's library. This alteration proved to be an additional clog to my business (which besides was daily increasing on me) for I lost by this means my workroom and found it very difficult to keep the necessary order among the MSS.[14]

In 1840, over 50 years after the initial upheaval, Herschel recorded how upsetting this period was to her daily routine, and how she felt it had adversely affected her career. The last 24 years of her residence in England, between 1798 and 1822, Herschel noted, required her to change habitation seven times, 'which was always attendant with useless expenses'. More exasperating, however, she stated emphatically, was the destruction of 'what was still more precious, <u>loss of time</u>, which to this present moment I cannot help to grieve at when thinking on the confused and unfinished state in which the Register of Sweeps were left'.[15] Once again interrupted by others in her pursuit of a profession, Caroline Herschel still suffered the searing pain of being thwarted in her ambitions over half a century after the event.

And yet to others, very little had changed in the

lives of the Herschel siblings. Guests and visitors recorded their observations of the working relationship between William and Caroline, often in awe of the apparently seamless connection between the two. Frances Burney met the Herschels again in September 1787, desirous to reacquaint herself with Caroline, whom she 'wished to see very much, for her great celebrity in her Brother's Science', marvelling, in a letter to her sister Susanna Phillips, at the method of communication between the two:

> Their manner of working together is most ingenious and curious. While he makes his observations without Doors, he has a method of communicating them to his Sister so immediately, that she can instantly commit them to paper, with the precise moment in which they are made. By this means, he loses not a Minute, when there is any thing particularly worth observing, in writing it down, but can still proceed, yet still have his accounts and calculations exact.[16]

In 1799, a Frenchman, Barthélemi Faujas de St-Fond, who travelled Britain in order to ascertain the state of the arts and sciences in the country, rhapsodised about the immense contribution of the Herschels to science. This 'fraternal communication', he enthused,

> applied to a sublime but abstruse science, this constancy of study, during successive nights, employed in great, difficult observations, afford pleasing examples of the love of knowledge, and are calculated to excite an enthusiasm for the sciences, since they present themselves under an aspect so amiable and so interesting.
>
> ... That man must be born with a very great indifference for the sciences, who is not affected by this delightful accord, and who feels not a desire that the same harmony should reign among all those who have the happiness to cultivate them. How much more rapid would their progress then be![17]

St-Fond noted that, by means of a string which connected brother to sister, William Herschel was able to attract Caroline's attention. She would open the window and then he would ask her for information. After consulting the tables before her, Herschel would reply 'brother, search near the star *Gamma, Orion*, or any other constellation which she has occasion to name. She then shuts the window and returns to her employment.'[18] The intellectual traveller also gave an account of Caroline Herschel's role in the partnership, from her own mouth:

'My brother', said Miss Caroline Herschel, 'has been studying these 2 hours; I do all I can to assist him here. That pendulum marks the time, and this instrument, the index of which communicates by strings with his telescopes, informs me, by signs which we have agreed upon, of whatever he observes. I mark upon that large chart the stars which he enumerates, or discovers in particular constellations, or even in the most distant part of the sky.'[19]

As St-Fond's account revealed, visitors continued to notice the large part the 'astronomical assistant' played in the discoveries of her brother, as well as praising her own abilities.

The continuing construction of the 40-foot telescope encouraged more and more interested parties to come to Slough. One double page of the Visitors' Book for 1788 noted visits from royalty, aristocracy (including the Duchess of Devonshire) and several professional scientists. In August and September, for example, the Herschels hosted Lalande, Scottish academic John Playfair and Professor Meyer of Göttingen.[20] The elderly Herschel remembered anecdotes of the great and the good with wicked humour. Reflecting on French visitors before the Revolution of 1789, Herschel recollected the courtier the Princesse de Lamballe, who 'about a fourthight af[t]er' she had visited Slough became a victim of the guillotine: 'Her Head was off!'[21] In another letter written in 1840, Herschel noted the visit in August 1787 of the King, Queen, members of the royal family and the Archbishop of Canterbury.

The telescope tube was just over 40 feet long and nearly five feet in diameter, and it had been constructed by William Herschel in a large barn near Upton church, about a quarter of a mile from his house. It was made of large pieces of sheet-iron seamed together and then strengthened by iron hoops and longitudinal bars.[22] Unsurprisingly, this giant structure became quite a novelty attraction. Before the telescope was mounted, its huge tube was a veritable magnet for either the professionally or the fashionably intrigued.

One Anecdote of the old Tube (if you have not heard of it) I must give you. Before the Optical parts were finished, many Visiters had the curiosity to walk through it, among the rest King Georg the III and the Bishop of Canterbury – following the King, finding it difficult to follow; the King turned to give him the hand saying 'Come My Lord Bishop I will shew you the way to Heaven'.[23]

Further events drew the curious. Foreshadowing the banquet held underground in 1827 – when another remarkable feat, Marc and Isambard Kingdom Brunel's Thames Tunnel, was being constructed – an impromptu concert was mounted in the tube:

Perhaps you may have heard that in the early part of its existence God save the King was sung in it by the whole company who got up from dinner and went into the Tube among the rest 2 Miss Stows, the one a famous Piano-forte player and some of the Griesbachs who accompanyd on the Oboe or any Instrument they could get hold of.

'And I', the ageing Caroline remembered with a sigh, 'was one of the nimblest and foremost to get in and out of the Tube'.[24]

But the ease with which Caroline moved in and out of the tube was not matched by its difficult and lengthy construction. The first mirror, which had been cast in 1785, was too thin in the centre. This William Herschel blamed on the 'mismanagement of the person who cast it'.[25] But it was decided that the mirror would be used, and it was set in an iron ring for grinding and polishing. Twelve men were required to position the mirror over a convex iron tool and turn the ring's handles so that the abrasive movement would create the curvature required. While the iron tool was covered in pitch, the mirror was suspended by a crane, dipped into a tub of hot water and then placed on the polisher until it achieved the correct curvature. The pitch then had a deep cross-gutter cut into it to allow polishing to begin. The sheer size of the mirror created problems from the outset, because it did not allow the delicacy with which William and his sister had previously polished smaller mirrors. But the polish was successful and, 'though far from perfect', the mirror allowed William a good view of the extremely bright nebula in the belt of Orion.[26] Two more mirrors were created; the second cracked when cooling, but the third was successful, even though it had not been brought to the perfect condition William Herschel expected. In constructing his own machine by the summer of 1789 for 48-inch disks, William managed to improve the mirror. When the telescope was mounted, its weight and that of the observing galleries was supported by a 50-foot triangular latticework of poles and ladders. Based upon an octagonal platform and rotated by sets of rollers on two brickwork circles, the tube could be turned by fixing a rope around a grounded pole from a windlass to one corner of the platform. By a system of ropes and pulleys, the upper end of the tube was suspended between two sets of ladders and attached to a high transverse beam 40 feet above the ground. The observer on the platform could move the tube laterally, or a winch would position the tube vertically. The astronomical assistant – Caroline, of course – sat in a hut or observatory on the ground platform.[27]

The 40-foot reflector was never the astronomical success for which William Herschel had so fervently hoped. It was far too cumbersome and, due to the alloy used, the mirror tarnished quickly. Even the wooden framework suffered from rot.[28] In time it became more of a popular attraction than a working instrument, and indeed, as if to mirror this transition, the first Ordnance Survey map of the area (1830) listed the telescope as a notable feature of the landscape. But the initial interest it had provoked in the public brought those eager to meet the famous Herschels. Now that Caroline, in particular, had made a name for herself, others sought to assess her character in order to examine her especial qualities. What many found was not really what they had expected; a curious combination of features emerged from their discoveries. Despite noticing her 'indefatigable' devotion to her brother, Mrs Papendiek, as a friend of Mary Herschel, was clearly not that impressed with the female astronomer and suggested rather cruelly that Caroline Herschel was 'by no means prepossessing'. From this ambiguous focus, which could have been either upon appearance or personality, Papendiek redeemed herself by claiming that she was 'a most excellent, kind-hearted creature, and though not a young woman of brilliant talents, yet one of unremitting perseverance, and of natural cleverness'.[29] Sophie von La Roche was far kinder, and concentrated on innate qualities rather than looks. She found Herschel 'all gentleness, sensibility and humility', and both siblings were 'through their close contact with the constellations ... raised above all artificiality and conceit'.[30] In similar fashion to La Roche, Frances Burney found Herschel

> very little, very gentle, very modest, and very ingenuous: and her manners are those of a Person unhackneyed and unawed by the World, yet desirous to meet, and return its smiles. I love not the philosophy that braves it: this Brother and Sister seem gratified with its favour, at the same time that their own pursuit is all-sufficient to them without it.[31]

This quality of both Herschel siblings' 'self-sufficiency' was, or so contemporary philosophical theory would put it, precisely that of the scientist. For all her apparent modesty, to her contemporaries Caroline Herschel appeared to embody exactly those detached feelings necessary for the pursuit of science.

Caroline's apparent indifference to her own discoveries has led previous historians of the Herschel family to conclude that she was simply not interested in the science which she pursued solely because her brother told her to do so. Indeed, Herschel sometimes colluded in this assessment of her reputation, if accolades afforded her too much credit to the detriment of her brother's original achievements. The reader of Herschel's manuscripts may look vainly 'for any hint of interest beyond the mere act of discovery: in the orbits of her comets and in the possibility of previous and future apparitions, to say nothing of the physical nature of comets, and their role in the economy of nature'. Indeed: 'Once she found a comet she proudly handed it over to astronomers, her job done.'[32] Similarly, others have discussed the implications of the lack of direct female involvement with the sciences in the late eighteenth and early nineteenth centuries.[33] Feminist historians have argued that science itself is inextricably linked to 'popular hallmarks of masculinity', such as objectivity, rationality, truth, progress, exploration and power.[34] The skills involved in original discovery were becoming more and more important as the individual sciences began to splinter off from a more broad-based natural philosophy. It is, therefore, precisely the more feminine area of explication, of contemporary scientific ideas distilled for a popular market, where women were able to succeed in this period. This was a role that could be labelled as scientific 'mediatrix'.[35] In a sense, therefore, masculine scientific theorising was the mid-point in a chain which extended from discovery to universal explanation. As a woman who had actually discovered astronomical objects, Caroline Herschel, in contrast to her female contemporaries whose only link to science was through the popularisation of ideas, was always two steps ahead.

The calm, self-sufficient demeanour which both Herschels displayed characterised them instantly as intimately involved in the pursuit of the sciences. In 1759, the philosopher and political economist Adam Smith published the popular and influential *Theory of Moral Sentiments*, which commented precisely on this issue. In a section examining how certain types of people reacted to public opinion, Smith noted the differences between those who practised the arts and those who practised the sciences. A poet, for example, claimed Smith, delighted in the approbation of others and reacted severely to their disdain. The concern of the poet for universal applause stemmed from the unique sensibility necessary for obtaining success in poetry and the fine arts, and the resulting uncertainty that one can ever attain true beauty in one's productions. Conversely, mathematicians and natural philosophers or scientists, who are assured about the importance and the truth of their discoveries, often manifest indifference about their public reception. This sense of security and tranquillity ensured that mathematicians and scientists were calmer, more open about their finds, and less factional than poets or those who indulged in the art of fine writing:

> Mathematicians and natural philosophers, from their independency upon the public opinion, have little temptation to form themselves into factions and cabals, either for the support of their reputation, or for the depression of that of their rivals. They are almost always men of the most amiable simplicity of manners, who live in good harmony with one another, are the friends of one another's reputation, enter into no intrigue in order to secure the public applause, but are pleased when their works are approved of, without being either much vexed or very angry when they are neglected.[36]

Caroline Herschel was secure in the knowledge that she was the original discoverer of astronomical objects. Her measured reaction to the public as well as to her own discoveries, as noticed by her contemporaries, in fact would have made her, in the eyes of one of the century's most important and influential philosophers, an ideal scientist.

Indeed, another vital contribution made by Caroline Herschel to the pursuit of astronomy was finally published in 1798, and represented sixteen years of dedication to the science she had learned from scratch. Already in this decade she had been the first woman to publish notices of her original discoveries in the Royal Society's prestigious journal, the *Philosophical Transactions*. Her scientific work was published and recognised under her own name – a unique achievement for a woman in this period. William Herschel had hinted in advance in his own published papers that his sister was about to produce something of great value to astronomy. In the *Description of a Forty Feet Reflecting Telescope*, published in 1795, he noted that

> no catalogue of stars in zones had ever been published; I therefore gave a pattern to my inde-fatigable assistant, Caroline Herschel, who brought

all the British catalogue into zones of 1° each, from the 45th degree of north polar distance down to the horizon, and reduced the night ascension of the stars in it to time, in order to facilitate observations by the clock. This catalogue was afterwards completed from the same degree up to the pole in zones of 5° each; and the variation in night ascension from one deg of change in longitude, was also reduced to time, for every star in the catalogue. To this were added completed tables for carrying back present observations to the time of that catalogue; which method I preferred to bringing the stars it contains forward to the present time, on account of conforming with the construction of the Atlas Coelestis, which was of great service.[37]

This essential aid to astronomical 'facilitation' was further mentioned in 1797 in William Herschel's *A third catalogue of the comparative brightness of the stars*, in which he proved to be his sister's mouthpiece, claiming that this 'considerable catalogue', which would add between 500 and 600 stars overlooked when Flamsteed's British Catalogue was framed, 'is already drawn up and nearly finished by Miss Herschel, who is in hopes that it may prove a valuable acquisition to astronomers'.[38] The 'usefulness' of the achievement, the necessity of making herself useful, to herself and to others, was, of course, a lifelong desire of Caroline Herschel.

As William Herschel stated, the catalogue containing the almost 3,000 stars observed by the first Astronomer Royal, John Flamsteed, had been published in the second and third volumes of his *Historia coelestis Britannica*. What became known as the British Catalogue was published in 1725, six years after Flamsteed's death, and became the authority on the stars.[39] But William Herschel did not find the work as authoritative or reliable as he had hoped. As the observations were published in two separate volumes, constant cross referencing between the texts was necessary, and there were also errors and omissions in the work. This was particularly noticeable during observations, when stars which had been recorded in the Flamsteed volume seemed to have disappeared. However, the omission was, in fact, due to printer errors or an incorrect transcription of the original, posthumous production.[40] Too wrapped up in his own work, William encouraged his sister to carry out the laborious calculations and amend the catalogue. When it was published, the *Catalogue of Stars, taken from Mr Flamsteed's Observations contained in the second volume of the Historia Coelestis, and not inserted in the British Catalogue*, contained an additional *Index to point out every observation in that volume belonging to the stars of the British Catalogue, as well as a collection of errata that should be noticed in the same volume.* The value and magnitude of the work was recognised by the fact that the Royal Society, persuaded of its fundamental necessity by Maskelyne, funded all the costs of publication.[41] Five hundred and sixty-one stars had been found by Caroline Herschel which had been overlooked in the original.[42] The amateur astronomer Edward Pigott, who had known the Herschels in Bath, summed up the importance of her achievement in a letter to Caroline of 30 April 1799:

> Were Flamsteed alive, how cordially would he thank you for thus rendering the labours of his life so much more useful and acceptable to posterity, for he surely little thought that his great

work required to be elucidated by an additional folio volume of explanations, errata, and indexes, the advantages of which, by their excellence and accuracy, must every day be more and more acknowledged, and future astronomers, as well as those of the present times will doubtless often be conscious of the merit and obligation you are entitled to.[43]

The astronomers Francis Baily and Nevil Maskelyne expressed some doubts over the organisation of the catalogue, which, for Baily, was exacerbated by Herschel's arrangement of the observations in the same order as Flamsteed instead of their right ascension. But he praised the juxtaposition of a known with an inedited star, which, with the difference of their right ascension and declination, allowed the possibility of determining the location of the latter.[44] Maskelyne's objection was that the list of omitted stars should have been presented in the sequence to which they appeared to observers rather than by constellation. Caroline Herschel rose to the challenge and prepared the list in the form Maskelyne wanted.[45] After all, it was Maskelyne's suggestion in the first place that the work be published, a recommendation that, she acknowledged, flattered the 'vanity' of the ambitious Caroline Herschel not a little.[46]

The *Catalogue* has been described as 'immensely useful, yet it required nothing more than perseverance and an infinite capacity for taking pains'.[47] Such an assessment does, in fact, remove an incredible achievement from Caroline Herschel. She had provided the scientific community with a work that facilitated all their own discoveries. She had helped to improve upon a catalogue which had previously been surrounded with an almost religious aura, and won accolades both for her own reputation and for creating a work which would assist the reputation of others. As the editor of William Herschel's scientific papers, J.L.E. Dreyer, recognised, Caroline had attained far more than simple acclaim for her clerical skills. That she was able to deal so efficiently and correctly with the abstract nature of her task should make her achievement here all the more remarkable. And all this in spite of her sex, which was considered to disable women from the pursuit of such sciences in the opinion of some eighteenth-century thinkers:

Caroline Herschel's perseverance in doing the enormous amount of copying required of her was equalled by her carefulness and accuracy in performing her tasks; the books and slips written in her clear and most legible hand seem completely free from slips, either clerical or arithmetical. In fact the only mistakes she ever made in her work seem to have been two or three slips of 1° in Polar Distance, in the Zone Catalogue of her Brother's nebulae which she made in her old age.[48]

Surely an amazing feat for anyone, let alone a middle-aged, ill-educated, autodidactic female from Hanover. Most importantly, Herschel recognised her own achievement in carrying out the alterations to the Flamsteed catalogue. While she was pleased that her brother's burden had been lightened, it was delight at the feeding of her vanity and ambition which stood out prominently from her letter to Maskelyne noted above.

After the hustle and bustle of the 1790s, life became more sedate, if not less busy for Caroline Herschel. In 1797, just after the discovery of what was to prove her final comet, she moved house. This

appears to have been a decision that Herschel regret-
ted. The 'Day-Book' that she kept from 1797 to 1821
began:

> 1797, in October I went to lodge and board with
> one of my brother's workmen (Sprat), whose wife
> was to attend on me. My telescopes on the roof,
> to which I was to have occasional access, as also
> to the room with the sweeping and observing
> apparatus, remained in its former order, where I
> most days spent some hours in preparing work
> to go on with at my lodging.[49]

This suggestion of restricted access through only
occasional visits to her own instruments perhaps
makes clear why Herschel had no more original discoveries after 1797. Living at some distance from
her telescope meant lengthy preparations for visiting at night. When she lived in Slough High Street
she would be compelled to watch the condition of a
cloudy sky for a potential clearing. If visibility became
sufficient, Herschel would wake her landlord's son
by knocking on his bedroom wall. The boy would
then accompany Herschel with a lantern over to the
telescope and a waiting William.[50] Other factors also

intervened. In the first decade of the nineteenth century, Herschel became increasingly fearful that she was losing her sight. This was, of course, a terrifying prospect for a woman whose keen eyesight had aided her substantially in her astronomical activities. Herschel, unsurprisingly, experienced 'dreadful apprehension' about her condition, which was exacerbated by the slowness of her doctor in pronouncing her fears to be for nothing.[51] For a woman who supposedly paid little attention to the sociability of the world around her, 'supremely indifferent to what was going on in the world outside her own little circle', Herschel's sparse entries in her Day-Book reveal an active and fulfilling round of visiting and being visited.[52] In 1802, she was reacquainted with the friendly child she had met at her millinery lessons in Hanover in 1767, who, to Herschel's great surprise, delightedly renewed the friendship. Mrs Beckedorff was now a member of the royal household, and soon acquainted her old friend with several other members. Herschel was favoured particularly by the Princesses Sophia and Amelia, who enjoyed her conversation and often requested her company.[53] Herschel met Frances Burney again, now Madame D'Arblay, twice between

May 1813 and December 1814, and spent 'very pleasant' evenings with her old acquaintance.[54] Only William's failing health could spoil an invitation to a fête at Frogmore in July 1817:

> [A]lmost as soon as the Royal party sat down to dinner I was obliged to go home with my brother, after having twice been honoured by the notice and conversation of the Queen and Regent, &c., &c. He found himself too feeble to remain in company. It was said that there were above 2000 persons invited.[55]

The resulting early departure rankled with Herschel, who was evidently enjoying all the attention she received on the occasion.

Yet notice of Caroline Herschel's achievements was also to be received in published writings. While appearing as both an assistant and an astronomer in her own right in *British Public Characters of 1798*, Herschel later featured in a survey of distinguished European females proficient in the physical sciences. The director of the Paris Observatory, Joseph Jérôme de Lalande, one of Herschel's greatest admirers, wrote a *Ladies' Astronomy* which lauded the intellectual capacity of famous scientific females from all countries and all times, ranging from Hypatia to the more recent Emilie du Châtelet, who translated and published the works of Newton in France; and the Countess of Purinina, founder of an observatory in Poland, Madame du Piery, the first woman to teach astronomy in Paris, and, of course, Caroline Herschel, 'whose proficiency in the science is so well known, has discovered five [sic] comets'.[56] Perhaps one of the most famous women scientists of eighteenth-century Europe was the highly fêted Italian lecturer Laura Bassi, who was awarded a chair in experimental physics at the Bologna Academy for the Institute for Sciences two years before her death in 1776 and had been paid, from 1760, 400 lire more than her most important male colleagues.[57] Not only were women capable of observing the heavens and displaying 'great perseverance' in doing so, they were also sufficiently talented to discover as well as passively observe. The 'spirit of inquiry' was alive and thriving within female minds.[58] Herschel's discoveries not only placed her within a contemporary hall of fame but also revealed her as a female scientific exemplar.

However, further accolades could not eclipse the

increasing worries that Caroline felt for her ageing brother. William's health deteriorated rapidly in the late 1810s and he began to require his sister's help in organising his papers. He had a literal gift to posterity in the form of his brilliant mathematician son John, who had been born in 1792, and whom William had encouraged to abandon an academic career to learn the family trade of astronomy. At the age of 71, in June 1821, Caroline Herschel acted as John's assistant, recording his first sweeps, as she had, for so many years, his father's.[59] The aunt was exceptionally fond of the nephew, who was later frequently mistaken for her son, and towards whom she indeed expressed 'Motherly feelings'.[60] But Caroline Herschel was increasingly depressed, and this feeling of foreboding was reflected in the final entry of her Day-Book in October 1821: 'Here closed my Day-book, for one day passed like another, except that I, from my daily calls, returned to my solitary and cheerless home with increased anxiety for each following day.'[61] The old terror about what would happen to her in a future without William and without employment clearly possessed a woman who had now reached her seventies and seemed at her most vulnerable. It was now that she began composing the start of the *First Autobiography* for Dietrich, and bitter memories of the past mingled with hopelessness for the future. William Herschel's death on 22 August 1822, at the age of 83, only deepened his sister's anxieties.

Even before the death of her beloved older brother, Herschel had been contemplating her future. Sophia, Jacob, William and Alexander were dead, and now Dietrich was the only Herschel sibling left for Caroline to turn to in her old age. In fact, Herschel would outlive her younger brother by 23 years. Dietrich had become increasingly concerned about his finances and the dire straits Hanover once again found itself in during the Napoleonic Wars. He visited William and Caroline in 1808, and eventually stayed for four years. Herschel expressed her disdain towards a brother who treated her like the compliant elder sister he had once had. Caroline was angry at the treatment meted out to her as the only female of the family:

From the hour of Dietrich's arrival in England to that of his departure, which was not till nearly four years after, I had not a day's respite from accumulated trouble and anxiety, for he came ruined in health, spirit, and fortune, and according to the old Hanoverian custom, I was the only one from whom all domestic comforts were expected. I hope I have acquitted myself to everybody's satisfaction, for I never neglected my eldest brother's business, and the time I bestowed on Dietrich was taken entirely from my sleep or from what is generally allowed for meals, which were mostly taken running, or sometimes forgotten entirely. But why think of it now!

With Dietrich's presence, Herschel's unexpected return to the 'old customs' reminded her of the horror of the past.[62] And yet Herschel still had a soft spot for her younger brother. The £10 which had been added to her salary by William and Mary Herschel in 1803 was given to Dietrich during his four-year stay.[63] She would later hand over to him her savings of £500, believing, clearly, that her future would lie with him.[64]

And so it did for a time, because on 10 October 1822, Caroline Herschel left Slough to return with a very heavy heart to her native land. Before she left, she may have noticed a generous mention of her own powers in the *Gentleman's Magazine* obituary of William. The periodical had concluded that:

> In these observations, and the laborious calculations into which they led, he was assisted throughout by his excellent sister, Miss Caroline Herschel, whose indefatigable and unhesitating devotion in the performance of a task usually deemed incompatible with female habits, surpasses all eulogium.[65]

William's will awarded Caroline an annuity of £100, which was in stark contrast to the £2,000 left to the impecunious Dietrich.[66] But at least it would make her reasonably comfortable in her twilight years, which she was convinced would soon end. The move to Hanover, however, was something she could never look upon without pain and severe distaste. This was expressed throughout the rest of her life, but perhaps most succinctly registered in a muddled and angry letter written in 1840 to John Herschel's wife Margaret: 'O! why did I leave England?'[67] This sense of regret at missing out on the advantages of an English life never left Caroline Herschel.

But this is in no way to suggest that Herschel did

not make the most of her time in Hanover. Her letters from 1822 until the end of her life are maudlin and self-pitying but also sharp, intelligent and touching, and they contain what most historians of the Herschels deny that Caroline possessed – a witty and wicked sense of humour. In fact, she was anything but 'determined to be miserable' or 'an embittered, prickly, solitary woman who shunned friendly advances and perceived the world as a hostile place' in the correspondence she carried out with her family, her brother's widow and then with her nephew and his wife, and other correspondents.[68] Despite the continuing laments for her English life, constant denial of her Hanoverian birthright and typically elderly comments about her failing health, Herschel largely enjoyed the years she had left. She went to 'Playhouses and concerts' twice a week and enjoyed the notice English visitors in particular gave her. Her presence at any public event was a momentous occasion, she joked: '[I] amuse myself sometimes with having my vanity tickled by the notice which is taken of my being or not being present.'[69] Visits to a circle of 'Learned Lady's' much amused Herschel, and she laughed at the fact that she increased her reputation by remaining enigmatic and exploiting the expected frailties of her age. 'But I am imitating Robinson Cruso', she informed her nephew, 'who kept up his consequence by keeping out of sight as much as possible when he acted the Governour, and when they want to know anything of me I say I cannot tell!'[70] Paying a visit to his aunt in 1832, John Herschel noted that the 72-year-old Caroline moved around Hanover on the 'full trot', singing, dancing, running about, and skipping up flights of stairs as 'wonderfully fresh' as people 'not a fourth of her age'.[71] Far from following her brother rapidly to the grave, Caroline Herschel would live until 1848, her desire to be 'useful' to herself and to others undiminished in extreme old age. And, far from forgetting her scientific past, Herschel continued to maintain more than a passing interest in astronomical affairs, and eagerly followed journal and newspaper articles about her brother and herself, frequently correcting interpretations of events. In a way, the letters from this period help to create the reputation of Caroline Herschel as the devoted, indefatigable woman who sacrificed her life for her brother. But it is vital not to suppress the fact that this correspondence also revealed that Herschel thought highly of her reputation and believed the awards she received in the 1820s and beyond for her contribution to astronomy were very well deserved, if far too late to be fully appreciated.

As if to compensate for the loss of her scientific occupations, Caroline Herschel began work within a few months. She appeared lost and unsettled without work to occupy her time. In a letter of December 1822 to her nephew, she lamented that constant visitors disturbed her, she could not 'get my books and papers in any order', and the local architecture made observing the night skies impossible: 'And at the heavens is no getting at, for the heigh roofs of the opposite houses.'[72] It is clear that Caroline desired to sweep the skies, even when she was not being asked to do so by William or by John. She certainly had enough impetus to seek out astronomical objects by herself. Her interest was certainly greater than historians of her family have assumed. But, unable to view the heavens, other ideas preoccupied her. Since the Herschel siblings had begun sweeping for nebulae and clusters of stars when they moved to Datchet in 1782, they had dramatically increased the number of known nebulae from about 100 to 2,500.[73] Herschel's work on a catalogue had been in progress since the 1780s, but it was only at the beginning of the new century that she had calculated the position of stars which appeared in the sweeps of the year 1800 and had arranged them into zones. This made the task of sweeping far more manageable. Now, as William had noted the position of nebulae with reference to one of the stars in his numerous catalogues, Caroline had the laborious task of producing a catalogue of nebulae which would match that of the stars.[74] She grasped at this challenge with undisguised relish. In August 1823, Herschel wrote to her nephew John, willing herself to live long enough to carry out the task he had set her: 'I have not above 6 hours tolerable ease out of the 24, and not one hour sound sleep, and yet I wish to live a little longer; that I might make you a more correct Catalogue of Nebulae the 2500. which is not even begon. but I hope to be able to make it my next winter's amusement.'[75]

A year later, in August 1824, she wrote to another correspondent, Miss Baldwin, a relation of William's widow, Mary, explaining her slowness of response:

I left your letter thus long unanswered, because I would not make a break in the work in which I have for these last 3 months been engaged (the Catalogue of which I wrote in my last) which I wished to have ready for my Nephews inspection, in case I should see him. But if it should turn out otherwise, then, it must be a subject for our future correspondence.

Still, as I cannot do much at a time, I wish yet to live a little longer, for it is with me, as I suppose it is with every body; we always have something yet to do.[76]

For Caroline, work was still everything. John Herschel recognised the strength of the woman who had given all for so many years to so many people. In December 1824, in a draft of a letter to her, he wrote: 'The Sacrifices you have individually made for your family are above all praise.'[77] Only a month later, his aunt informed him that work was continuing apace: 'I am now writing out the Catalogue of Neb. and am at Zone 30° and hope to finish it for the Easter Messenger.'[78] And on 7 March 1825, Caroline made a succinct announcement: 'I am ready with the Catalogue of

Nebulae.'[79] Less than three weeks later, Herschel emphasised her achievement in completing such a task at the age of almost 75:

> I hope the M.S. Catalogue of Nebulae and that of the Stars, which I have observed in the Series of Sweeps; along with the 8 Volumns from which they have been drawn out, will not unfrequently be of use to you.
>
> I cannot make another Copy and this would besides introduce blunders; of which I hope there will not many be found, or at least none but what will be easily seen to be not only a slip of the pen.[80]

Priding herself on her accuracy and discipline, Herschel simultaneously drew attention to the laborious but also 'useful' nature of her work; qualities she had prized all her life and which were now brought together in this 'major personal contribution to astronomy'.[81]

Far from indifferent to the astronomical world, Caroline desired to be very much a part of current events. In a long passage from a letter to John, she wrote the following:

I cannot express my thanks sufficiently to you for thinking me worthy of forming any judgment of your Astronomical proceedings, and am only sorry that I cannot recal the health, eye sight and vigor I was blessed with 20 or 30 years ago; for nothing else is wanting (and that is all) for my coming by the first steam Boat, to offer you the same assistance (when Sweeping); as by your Father's instruction I had been enabled to afford him; For, an Observer at your 20 feet when Sweeping wants nothing but a being that can and will execute [the] commands with the quickness of lightening, for you will have seen that in many Sweeps 6 or twice 6 objects have been secured and described within the space of one minute of time.

I cannot think that any Catalogue but the M.S. one in Zones (which was only intended for your own use) would facilitate the Reviewing of the Nebulae. You will see in the Register of Fl[amsteed] time a curved line U which denotes that the milky way is in those places, and if you see an L. and find a cl[uster] of stars there about, I shall claim as one of those I mentioned in my letter to you. It was the assistant's business to give notice when such marks or any Nebulae in the lapping over of the Sweep either above or below were within reach; by making the workMan go a few turns higher or lower. NB. no more than is convenient without deranging the present Sweep. But I am forgetting myself and fear I am tiring you unnecessarily … I am quite impatient to see what you have to say about the Parallax of the fixt stars.[82]

Depressed at the loss of her once so sharp senses, Herschel here sought to instruct her nephew in the art of astronomical sweeping and revealed an awareness of the importance of her catalogue. Amusingly, she feared 'tiring' the man over 40 years her junior with her 'impatience' to hear more about his scientific theories. When John Herschel decided to complete his father's analysis of the heavens by mapping the southern hemisphere, Caroline cried, touchingly, and with not a little jealousy: 'Ja! if I was 30 or 40 years junger and culd go too? in Gottes namen!'[83] If only it were possible to 'shake off some 30 years from my shoulders that I might accompany you on your voyage', affording assistance or managing her nephew's 'work-shops', as she had done for his father when he was away.[84] Not quite content to remain behind, Herschel accompanied her nephew to the Cape of Good Hope like a vagrant 'in idea'.[85] But soon she was interfering in his searches, remembering William's frustration with what appeared to be 'ein Loch im Himmel', a hole in the heavens, in the 'body of the Scorpion'. Caroline too had 'stop[ped] afterward at the same spot' and was 'unsatisfied'.[86] John had found 'beautiful globular clusters of stars', but this was not what his aunt was after. She wanted him to confirm what both she and her brother had witnessed, which he duly did.[87] Contact with the world of astronomy through the movements of her nephew, the 'astronomical part of letters' and the 'promis of future accounts of uncommon objects' made Herschel 'very happy'.[88]

If Herschel recognised the magnitude of her own achievement, so did the European scientific community. Founded in 1820, as part of the explosion of interest in the sciences in the early part of the nineteenth century, the Astronomical Society of London, which would later become the Royal Astronomical Society, was, as Caroline Herschel put it herself,

'Florish[ing]'.[89] In February 1828, the Society awarded Herschel its Gold Medal for her recently completed catalogue, *The Reduction and Arrangement
in the Form of a Catalogue, in Zones, of all the Star-
clusters and Nebulae observed by Sir W. Herschel in his
Sweeps*. Previous winners of the medal had included
the great astronomers of the day: James South and
John Herschel himself (both 1826) and Francis Baily
(1827). Although Caroline's work had been originally
intended for John Herschel's own sweeping, he had
resolved to make it 'useful to others'.[90] And the grateful 'others' expressed their thanks. As John was the
current President of the Society, he allowed the Vice-
President, James South, to make the speech which
awarded Herschel the medal. While South claimed
that Caroline Herschel's 'labours' were inextricably
bound up with the achievement of her brother, he
declared that delivering a 'eulogy' upon the memory
of William Herschel was not the purpose for which
he spoke. He stressed the usual aspects of Caroline's
indefatigable assistance:

Who participated in his toils? Who braved with
him the inclemency of the weather? Who shared

his privations? A female. Who was she? His sister. Miss Herschel it was who by *night* acted as his amanuensis: she it was who conveyed to paper his observations as they issued from his lips; she it was who noted the right ascensions and polar distances of the objects observed; she it was who, having past the night near the instrument, took the rough manuscripts to her cottage at the dawn of day and produced a fair copy of the night's work on the following morning; she it was who planned the labour of each succeeding night; she it was who reduced every observation, made every calculation; she it was who arranged everything in systematic order; and she it was who helped him to obtain his unperishable name.

But he also emphasised her qualities as an astronomer:

[H]er claims to our gratitude end not here; as an original observer she demands, and I am sure she has, our unfeigned thanks. Occasionally her immediate attendance during the observations could be dispensed with. Did she pass the night in repose? No such thing. Wherever her brother was, there you were sure to find her. A sweeper planted on the lawn became her object of amusement; but her amusements were of the higher order, and to them we stand indebted for the discovery of the comet of 1786, of the comet of 1788, of the comet of 1791, of the comet of 1793, and of the comet of 1795, since rendered familiar to us by the remarkable discovery of Encke. Many also of the nebulae contained in Sir W. Herschel's catalogues were detected by her during those hours of enjoyment. Indeed, in looking at the joint labours of these extraordinary personages, we scarcely know whether most to admire the intellectual power of the brother, or the unconquerable industry of the sister.[91]

While South seemed unable to grapple with the concept of a woman undertaking astronomical duties for anything other than pleasure, he did highlight her original discoveries, made all the more impressive, he led his audience to believe, because they were so 'leisured'.

Further awards and recognition followed in the 1830s. In 1835, the Royal Astronomical Society elected

Herschel to an honorary membership, along with Mary Somerville. Somerville was the translator of a complex mechanical treatise by Laplace, *The Mechanism of the Heavens* (1830), and, as author of original experiments in the 1820s and 1830s on sunlight and magnetism, the second female after Herschel to publish in the *Philosophical Transactions*.[92] In 1834, in a review of Somerville's *On the Connexion of the Physical Sciences*, William Whewell had proposed, for the first time in print, the label 'scientist'.[93] Both women had shown a great interest in each other's work, and Herschel lamented that she had not been able to meet Somerville before she left for Hanover; 'tell me a little more' about her, she asked eagerly of her nephew's wife.[94] Herschel was upset in 1832 when articles on Somerville seemed to disappear in the post (they eventually turned up in December of that year),[95] and she received in 1837 a copy of the *Connexion of the Physical Sciences* from the author, referring to the gift as a 'valuable present'.[96] In her letter thanking the Astronomical Society of London for the award of honorary membership, Herschel expressed it a 'great honor' to see her name joined with the 'much distinguished Mrs Somerville'.[97] For

Somerville, however, the honour was all hers. She was in awe of the older woman's 'talents', and wrote almost exactly the same of Herschel as Herschel had written of her: 'To be associated with so distinguished an astronomer was in itself an honour.'[98]

Less patronising than James South's address to the Society when Caroline Herschel won the Gold Medal, the Council reported that it had 'no small pleasure' in recommending Herschel and Somerville to the list of honorary members. The step was entirely consistent with 'propriety', as, 'in an astronomical point of view', there 'can be but one voice': 'the time is gone by when either feeling or prejudice, by whichever name it may be proper to call it, should be allowed to interfere with the payment of a well-earned tribute of respect.' Nor would such an acknowledgement to women be judged any 'less severely' than those awarded to men. Sex should not be an obstacle preventing women from achieving scientific recognition. The career of Caroline Herschel, along with that of Mary Somerville, effected a revolution in the public reward of female scientific practitioners. These were women, as the Council of the Astronomical Society put it, 'of whose astronomical knowledge, and of the utility of the

ends to which it has been applied, it is not necessary to recount the proofs'[99] – women, indeed, whose
astronomical achievements spoke for themselves.

Caroline Herschel's honours were increased when
she became an honorary member of the Royal Irish
Academy in 1838[100] and when, at the age of 96, in
1846, the King of Prussia awarded her the Gold
Medal for Science for sisterly assistance, but also, as
the scientist and explorer Alexander von Humboldt
wrote, for 'discoveries, observations, and laborious
calculations'.[101] In 1828, there had even been the
possibility of Caroline being awarded one of the
Royal Society medals but, due to her long period of
astronomical inactivity, it could not be awarded: 'the
rule limiting the time within which those medals
must be granted being precise, it could not be done
without a violation of principle.'[102] The increasingly
distinguished elderly lady viewed these awards with
not a little scepticism. Her reaction has typically
been employed by historians to illustrate her genuine
lack of interest in the pursuit of science if it did
not involve any mention of her brother's career.[103]
Admittedly, Herschel would not have been Herschel
if she did not refer to her canine status, her tool-like

employment by her brother.[104] When she received her first gold medal from the Astronomical Society she bewailed James South's eulogy: 'Whoever says too much of me! says too little of your dear Father! and can only cause me uneasiness.' '[T]hroughout my long spent life', she sighed, 'I have not been used, or had any desire of having public honours bestowed on me.'[105] However, when she first heard of the award, she proclaimed it a triumph, a 'great and unexpected honour of a Medal bestowed on me', and desired to know all about the history of the medal, whose head was on it, and if the impression was permanent.[106] Almost three weeks later, Herschel mentioned to her nephew that, after her death, she wanted the medal to be placed 'among those you have of your Father's and your own'.[107] This mark of Caroline Herschel's astronomical skill she felt as distinct and as worthy an achievement as those of her brother and nephew.

Herschel received the awards of honorary memberships in exactly the same way. After exclaiming 'Godd knows what for' when appointed to the Royal Astronomical Society, she made an almost identical pronouncement when presented with an invitation to join the Royal Irish Academy: 'I cannot help of caling every now and then out aloud to myself <u>what is *that* for</u>?'[108] But Herschel was unable to deny that the recognition was very welcome. Appended to a copy of the Royal Irish Academy's letter to inform her of her honorary membership, Herschel seemed delightfully lost for words: 'More I cannot say on this subject at present, only just so much; it has given me pleasure!'[109] The reason why she baulked at the awards was that they were too late for her fully to appreciate them. Rather than disinterest or fears of impropriety, Caroline Herschel was saddened when she recorded her achievements because she had been unable to match them in recent years. Old age and Hanover had prevented her from continuing with her astronomy: 'I think almost it is mocking me to look upon me as a Member of an Academy – I that have lived here these 18 years (NB against my will and intention) without finding as much as a single Comet.'[110] What had been called '<u>Promotion</u>' in the Hanoverian newspapers rankled with Herschel for one typically Herschelian reason. Now living a 'useless life', Caroline was embarrassed to receive accolades because she feared that, recently, unable to maintain her cometcatching reputation, she had not deserved them.[111]

Mentally alert to the very end, Herschel despised the body that increasingly collapsed upon her and left her with a 'crazy chattered constitution'.[112] John Herschel's brood of children sent her 'clever Riddles' when she was 94, and she could ask Margaret Herschel to say 'many pretty things' to them but also retort proudly: 'I will not trouble you for the Key; for stupid as I am at those things I can find the letters at the first reading.'[113] When Herschel discussed her career in extreme old age, she defended her profession. Dietrich Herschel might have claimed that catching a fly 'with a leg more than usual' was as 'sublime' a study as 'catching a Comet', but Herschel disagreed.[114] She could not belittle the science she had pursued with so much success, nor could she forget the importance of her own comets. Taking issue with what the Bremen physician and astronomer Heinrich Olbers had written incorrectly about her discoveries, Herschel cried indignantly: 'This puts me in mind of Olbers saying somewhere, I had discovered 5 Comets – who wanted him to give the No of my Comets when he knew them no better.'[115] Caroline would zealously defend her reputation; if she could not forget her brother's legacy, nor could she allow others to misrepresent her own deserved position as an original observer. As she wrote in the letter in which she described herself as merely a tool:

> but if too much is said in one place, let it pass; I have perhaps deserved it in another by perseverance and exertions beyond female strength. well done![116]

Caroline Herschel had set a precedent in the history of science. The various careers she tried to perfect throughout her life reveal far more about the lack of opportunities for women in the second half of the eighteenth and first half of the nineteenth centuries than about her own failures and inadequacies. That she achieved so much, including her ambitions for independence, was despite these social disabilities. When she died on 9 January 1848, just over two months before her 98th birthday, the British periodical the *Athenaeum* recorded that she had achieved 'unusual distinction' both inside and outside the scientific community. The obituary mentioned her 'indefatigable zeal, diligence, and singular accuracy of calculation', as well as listing all her celestial finds.[117] The tombstone epigraph composed by Herschel herself offered the same analysis, remarking on the two facets of her career: sisterly devotion and independent achievement, astronomical assistance and original discovery. But, noticeably, the proud female astronomer listed her discoveries first, as well as noting her recognition by scientific societies.[118] In her own writings and correspondence Herschel recognised the importance of her own reputation and the vital nature of independence. Identifying a new comet, she informed Margaret Herschel, in a curious mixture of English and German, was like the game one played in childhood, crying out competitively in order to receive an apple as a reward: 'And after all it is only like the children play Wer am ersten kick ruft, sol den appel haben, Wo ide then alle rufen kick!kick!kick! und so &c.'[119] In German, the verb to cry, *rufen*, has the same stem as *Ruf*, reputation. In an apt analogy, rather than belittling her achievements, Caroline Herschel asserted the primacy of her original discoveries, while others could do little but follow in the wake of her reputation.

Conclusion

That Caroline Herschel has been perceived since her death as nothing but a devoted servant to her great brother is partially due to her own insistence on her inferiority. But her posthumous reputation was also constructed by others who, selective in their quotations from her letters and journals, quashed the spirit which battled ambitiously to gain independence. Herschel might have been embittered about her lot, but she also revelled in the profit her own exertions brought her. Her correspondence and the testimonies of her contemporaries allow a more nuanced picture of how women could participate in the public life of the late eighteenth century before the professionalisation of the sciences gathered pace in the second half of the nineteenth century.

Until this point, the distinction between science as an acceptable pursuit and as something to be avoided was evidently a very fine one. This insufficiently obvious dichotomy was easily exploited, allowing women not only to pursue the sciences, but to excel in precisely the abstract forms in which the philosophers Nicolas Malebranche and Jean-Jacques Rousseau believed them unable to succeed because of their sex. The popularisation of the sciences in polite society in the eighteenth century was countered by an increasing recognition that areas of scientific expertise were professional occupations to be rewarded financially. While scientific experimentation could be carried out for the benefit of the public, it was best effected by those qualified to do so and in private. The lecture theatres of the Royal Institution exemplified this division. While fashionable metropolitan society would be entertained from 1799 in a specially constructed lecture theatre by the likes of Humphry Davy and Michael Faraday, the same lecturers would also carry out experiments for an elite scientific coterie downstairs in the bowels of the Institution.[1] The sciences were simultaneously a leisure activity and a very serious business, and women were able to participate in them during this time of flux. Women such as Emilie du Châtelet, Mary Somerville and Caroline Herschel were able to situate themselves deliberately within the heart of contemporary debate, forging links with the male European scientific elite who were shaping an understanding of earthly and celestial mechanics through complex formulae or original discovery. From an autodidactic intellectual background, ambition to succeed propelled them to stellar scientific heights as well as financial reward. As the playwright Joanna Baillie wrote to her friend Mary Somerville, recent female scientific achievements had 'done more to remove the light estimation in which the capacity of women is too often held than all that has been accomplished by the whole sisterhood of poetical damsels and novel-writing authors'.[2] In the words of the President of the Royal Society, Sir Joseph Banks, Caroline Herschel herself had 'advanc[ed] the science [she] cultivated with so much success', through perseverance, stoicism, but also, importantly, through ambition and a lifelong desire for independence.[3]

While institutionalisation eventually ensured that female amateurs found it more difficult to gain access to the scientific hierarchy due to their lack of tertiary-level training, the achievements of Caroline Herschel did not disappear. If, on the one hand, she was lauded by some as the dutifully devoted sibling, others stressed the independent successes of her remarkable career. Indeed, her recognition by members of the campaign for female suffrage in the last quarter of the nineteenth century reveals precisely how important Caroline Herschel's original discoveries were to those battling to secure women's rights. Supporters of women's educational right to participate in science, such as Emma Wallington and Lydia Becker, would employ the example of Caroline Herschel's 'stupendous astronomical labours' to fight their cause. Wallington even suggested that '[t]he splendid renown attached to Sir W. Herschel's name was largely due to his sister's superior intelligence, unremitting zeal, and systematic method of arrangement'.[4] While Wallington's male audience attacked her for providing only a list of exceptional women who had simply followed where men led and not advanced a step beyond, Wallington turned the argument back upon her interlocutors to counter that women did not receive the same educational encouragement as men to advance, and thus proved her point about the need to cultivate the female intellect. If women had not been able to reach the summit of a scientific discipline, it was due to the fact that the apex had never been presented as attainable for the majority of women. Unbarring the way could result in the success achieved by the likes of Herschel and Somerville. Although she may have been astounded to find herself employed as a feminist icon, it is possible to imagine that, secretly, Caroline Herschel would have been delighted to discover that her life and adventures were 'useful' to others.

Notes

INTRODUCTION

1   *Caroline Herschel's Autobiographies*, ed. Michael Hoskin (Cambridge: Science History Publications, 2003), *The First Autobiography*, pp. 89–90.

2   Joseph Jérôme de Lalande to Caroline Herschel, 23 December 1801, British Library Add 37203, Babbage Papers on Astronomy, 9.

3   Caroline Herschel to her niece, Lady Margaret Herschel, 4 June 1844, British Library Egerton MS 3762, Herschel Papers, 148–9.

4   Caroline Herschel to Lady Margaret Herschel, BL Eg 3762, Herschel Papers, 3 April 1844, 144–5; 4 September 1844, 154–5.

5   Caroline Herschel to John Herschel, BL Eg 3762, Herschel Papers, 1 May 1834, 1–2.

6   Agnes M. Clerke, *The Herschels and Modern Astronomy* (London: Cassell and Company, 1895), pp. 139–40.

7   Patricia Fara, *Pandora's Breeches: Women, Science and Power in the Enlightenment* (London: Pimlico, 2004), pp. 164–5.

8   Mary Wollstonecraft, *A Vindication of the Rights of Woman* (1792), in *The Works of Mary Wollstonecraft*, eds. Janet Todd and Marilyn Butler (London: Pickering and Chatto, 1989), p. 105.

9   Wollstonecraft, *Vindication*, p. 98. For more detail about women and ambition in eighteenth-century Europe, see Dena Goodman, *The Republic of Letters: A Cultural History of the French Enlightenment* (Ithaca and London: Cornell University Press, 1994).

10 Wollstonecraft, *Vindication*, p. 65; p. 237.

11 Caroline Herschel to John Herschel, 24 June 1823, British Library Egerton MS 3761, Herschel Papers, 16–17.

12 Mrs John Herschel, *Memoir and Correspondence of Caroline Herschel* (London: John Murray, 1876), p. 96.

13 Fara, *Pandora's Breeches*, p. 150.

14 Caroline Herschel to John Herschel, 24 December 1826, BL Eg 3761, Herschel Papers, 56–7.

15 Mrs John Herschel, *Memoir*, pp. 212–13.

CHAPTER I

1   Caroline Herschel to Margaret Herschel, 24 September 1838, BL Eg 3762, Herschel Papers, 32–3.

2   When Wilhelm and Carolina Herschel came to Britain, in 1764 and 1772 respectively, they became William and Caroline. Caroline Herschel later signs herself by this name, but refers to William as Wilhelm at times, and he calls her by the diminutive Lina.

3   *Caroline Herschel's Autobiographies*, ed. Hoskin, *The First Autobiography*, pp. 11–15. For the marriage date of Anna and Isaac, and therefore the fact of Sophia's illegitimate conception, see Michael Hoskin, *The*

Herschel Partnership as viewed by Caroline (Cambridge: Science History Publications, 2003), p. 7.

4   Caroline's unsociability as a child linked her closely to her father, as the trait seems to have been inherited from Isaac who, 'of a morose disposition', 'encouraged [Anna] to keep up a social intercourse among a few acquaintances'. *Caroline Herschel's Autobiographies*, ed. Hoskin, *The First Autobiography*, p. 32.

5   For more detail on the Seven Years War, see Uriel Dann, *Hanover and Great Britain 1740–1760: Diplomacy and Survival* (Leicester: Leicester University Press, 1991), and Olwen Hufton, *Europe: Privilege and Protest 1730–1789*, 2nd edition (Oxford: Blackwell, 2000).

6   See the more detailed account in Hoskin, *The Herschel Partnership*, p. 9.

7   *Caroline Herschel's Autobiographies*, ed. Hoskin, *The Second Autobiography*, p. 103.

8   *Caroline Herschel's Autobiographies*, ed. Hoskin, *The First Autobiography*, p. 26; *The Second Autobiography*, p. 103.

9   *Caroline Herschel's Autobiographies*, ed. Hoskin, *The Second Autobiography*, p. 102; n. 9, p. 102.

10 *Caroline Herschel's Autobiographies*, ed. Hoskin, *The Second Autobiography*, p. 110.

11 *Caroline Herschel's Autobiographies*, ed. Hoskin, *The Second Autobiography*, p. 107.

12 *Caroline Herschel's Autobiographies*, ed. Hoskin, *The First Autobiography*, p. 29.

13 Caroline Herschel to John Herschel, 29 November 1840, BL Eg 3762, Herschel Papers, 66–7; 4 May 1843, 126–7.

14 Caroline Herschel to Margaret Herschel, 9/10 January 1840, BL Eg 3762, Herschel Papers, 50–1.

15 Caroline Herschel to Margaret Herschel, 30 July 1838, BL Eg 3762, Herschel Papers, 26; letter to P. Stewart, 25 May 1835, 9–10.

16 Caroline Herschel to Margaret Herschel, 29 April 1835, BL Eg 3762, Herschel Papers, 7–8.

17 Marie-Claire Hoock-Demarle, *La femme au temps de Goethe, 1749–1832* (Paris: Stock/Laurence Pernoud, 1987), p. 40.

18 Hufton, *Europe: Privilege and Protest*, p. 16.

19 *Sophie in London* (1788), trans. Clare Williams (London: Jonathan Cape, 1933), pp. 216–17.

20 These European notions concerning Germany and the Germans are quoted from T.C.W. Blanning, *The Culture of Power and the Power of Culture: Old Regime Europe 1660–1789* (Oxford: Oxford University Press, 2002), pp. 241–3.

21 Hufton, *Europe: Privilege and Protest*, p. 103.

22 George expressed the former view a year before he came to the throne, the latter in 1762. Letters to the Earl of Bute, 5 August 1759, and 1762, quoted in Blanning, *The Culture of Power*, pp. 322–3.

23 *Caroline Herschel's Autobiographies*, ed. Hoskin, *The First Autobiography*, p. 28.

24 Hufton, *Europe: Privilege and Protest,* pp. 86–90.

25 There tends to be a divide between social and economic historians and music historians about the significance of the absent monarch. For the former point of view, see Thomas Bauman, 'Courts and Municipalities in North Germany', in Neal Zaslaw, ed., *The Classical Era: From the 1740s to the End of the Eighteenth Century* (Basingstoke: Macmillan, 1989), pp. 240–67; p. 247. For the latter opinion, see John Gagliardo, *Germany Under the Old Regime 1600–1790* (London and New York: Longman, 1991), p. 368.

26 Dann, *Hanover and Great Britain,* p. 127.

27 Gagliardo, *Germany Under the Old Regime,* p. 290.

28 Statistics from Olwen Hufton, *The Prospect Before Her: A History of Women in Western Europe. Volume I: 1500–1800* (London: Fontana Press, 1997), p. 18; p. 424.

29 Mary Wollstonecraft, *A Vindication of the Rights of Woman* (1792), in *The Works of Mary Wollstonecraft* (London: Pickering and Chatto, 1989), Vol. 5, p. 235.

30 *Queen of Science: Personal Recollections of Mary Somerville,* ed. Dorothy McMillan (Edinburgh: Canongate, 2001), p. 18; p. 20.

31 Jane Austen, *Emma* (1816), ed. Ronald Blythe (Harmondsworth: Penguin, 1985), p. 52; *Sense and Sensibility* (1811), ed. Tony Tanner (Harmondsworth: Penguin, 1986), p. 179.

32 Peter Petschauer, *The Education of Women in*

Eighteenth-Century Germany: New Directions from the German Female Perspective. Bending the Ivy (Lampeter: Edwin Mellen Press, 1989), p. 140; p. 112.

33   Hoskin assumes that because Herschel was re-educated by her brother in mathematics, she did not learn any at school. Not only does she not refer directly to a mathematical education for herself, she suggests that *all* the Herschel children were taught only how to read, write and observe their religion. William noted elsewhere that he was taught arithmetic as well. However, with educational reform in northern Germany gathering pace at the time that Herschel was in school, it is certainly possible that she may have had some knowledge of the most basic rules of mathematics. It should not be taken for granted that 'Caroline, as a girl, had not been taught arithmetic at school', *Caroline Herschel's Autobiographies, The First Autobiography*, n. 4, p. 50.

34   Petschauer, *The Education of Women*, p. 177.

35   Petschauer notes that Halle, Waldeck, Prussia, Braunschweig-Wolfenbüttel and Saxony taught girls arithmetic; *The Education of Women*, p. 110.

36   See Petschauer, *The Education of Women*, pp. 236–71, for more detailed information.

37   Mary Astell, *The Christian Religion as Profess'd by a Daughter of the Church of England in a Letter to the Right Honourable T.L., C.I.* (London: S.H. for R. Wilkin, 1705), p. 296.

38 Wollstonecraft, *Vindication*, p. 241; p. 263.

39 See Ann Shteir, *Cultivating Women, Cultivating Science: Flora's Daughters and Botany in England 1760–1860* (Baltimore, MD and London: Johns Hopkins University Press, 1996).

40 *Selections from The Female Spectator: Eliza Haywood*, ed. Patricia Meyer Spacks (New York and Oxford: Oxford University Press, 1999), Book XV, pp. 187–97; pp. 195–6.

41 Nicolas Malebranche, *Recherche de la vérité* (1674), (Paris: Galerie de la Sorbonne, 1991), pp. 232–3.

42 Jean-Jacques Rousseau, *Emile, Or On Education* (1762), ed. Allan Bloom (Harmondsworth: Penguin, 1991), p. 387.

43 *Selections from The Female Spectator*, p. 189.

44 Malebranche, *Recherche de la vérité*, p. 233.

45 *Caroline Herschel's Autobiographies*, ed. Hoskin, *The Second Autobiography*, p. 108.

46 Petschauer, *The Education of Women*, p. 397.

47 See the statistics in Hufton, *The Prospect Before Her*, p. 132.

48 *Caroline Herschel's Autobiographies*, ed. Hoskin, *The Second Autobiography*, p. 112.

49 *Caroline Herschel's Autobiographies*, ed. Hoskin, *The Second Autobiography*, p. 113.

50 *Caroline Herschel's Autobiographies*, ed. Hoskin, *The First Autobiography*, p. 47.

51 See chapter 13 of Isabel Rivers, *Lady Mary Wortley*

Montagu: Comet of the Enlightenment (Oxford: Oxford University Press, 1999), for a detailed account of the controversy in the 1720s over inoculation in Britain.

52   Statistics from the *Gentleman's Magazine*, 24 (1754), 601

53   *Caroline Herschel's Autobiographies*, ed. Hoskin, *The First Autobiography*, p. 22. Hoskin notes that 'This naturally reduced Caroline's chances of ever receiving an offer of marriage' (p. 22). But, as we see in the following paragraphs, the reasons for Herschel's lack of marriageable qualities are more complicated than this.

54   *Caroline Herschel's Autobiographies*, ed. Hoskin, *The First Autobiography*, p. 43.

55   Mary Wollstonecraft, *A Vindication of the Rights of Men* (1790), in *The Works of Mary Wollstonecraft*, eds Janet Todd and Marilyn Butler (London: Pickering and Chatto), Vol. 5, p. 22.

56   Germaine de Staël, *De l'Allemagne* (1813), intro. Simone Balayé (Paris: Garnier-Flammarion, 1968), 'Les Femmes', p. 66. My translation.

57   *Caroline Herschel's Autobiographies*, ed. Hoskin, *The Second Autobiography*, p. 111.

58   *Caroline Herschel's Autobiographies*, ed. Hoskin, *The Second Autobiography*, p. 112.

59   Petschauer, *The Education of Women*, p. 362. My translation.

60   Anon., *Female Rights vindicated; or the Equality of the Sexes Morally and Physically proved* (London: G. Burnet, 1758), p. 62.

61   Anon, *Female Rights vindicated*, pp. 56–8.

62   Anon., *Female Rights vindicated*, p. 49.

63   Petschauer, *The Education of Women*, p. 128.

64   *Caroline Herschel's Autobiographies*, ed. Hoskin, *The First Autobiography*, p. 22.

65   *Caroline Herschel's Autobiographies*, ed. Hoskin, *The Second Autobiography*, p. 111.

66   Caroline Herschel to Margaret Herschel, 24 September 1838, BL Eg 3762, Herschel Papers, 32–3.

67   *Caroline Herschel's Autobiographies*, ed. Hoskin, *The First Autobiography*, p. 39.

68   See the description of the eclipse in *The First Autobiography*, p. 36 and *The Second Autobiography*, p. 112.

69   *Caroline Herschel's Autobiographies*, ed. Hoskin, *The First Autobiography*, p. 24.

70   *Caroline Herschel's Autobiographies*, ed. Hoskin, *The First Autobiography*, p. 41.

71   *Caroline Herschel's Autobiographies*, ed. Hoskin, *The First Autobiography*, p. 37.

72   *Caroline Herschel's Autobiographies*, ed. Hoskin, *The First Autobiography*, p. 37. The reference to learning how to make masculine bags and sword knots before feminine caps and furbelows is to show indirectly that Caroline was encouraged solely to serve her brothers.

Also, being able to make the more feminine items was a skill which would have been prized outside the home and could potentially lead to employment.

73 *Caroline Herschel's Autobiographies*, ed. Hoskin, *The First Autobiography*, p. 41; *The Second Autobiography*, p. 114. The absence of any servant in the Herschel household, until at least 1769, ensures that we can locate their social status as well below that of the average bourgeois. In Marie-Claire Hoock-Demarle's *La femme au temps de Goethe, 1749–1832*, the suggestion is that, at this time, 'the average number of domestics in a bourgeois family home [was] from 3 to 5, with some having 6 to 8 and even more' (p. 56; my translation). Tellingly, when a servant was hired in September 1769, the Herschel family did not have the space to allow their new domestic her own room. Consequently, she was compelled to share Caroline Herschel's bed; *The First Autobiography*, p. 45.

74 *Caroline Herschel's Autobiographies*, ed. Hoskin, *The First Autobiography*, p. 34.

75 *Caroline Herschel's Autobiographies*, ed. Hoskin, *The First Autobiography*, p. 39; *The Second Autobiography*, p. 114.

76 *Caroline Herschel's Autobiographies*, ed. Hoskin, *The First Autobiography*, pp. 41–2; *The Second Autobiography*, p. 116.

77 *Caroline Herschel's Autobiographies*, ed. Hoskin, *The First Autobiography*, pp. 44–6.

78 Caroline Herschel to Margaret Herschel, 24 September 1838, BL Eg 3762, Herschel Papers, 32–3.

79 *Caroline Herschel's Autobiographies*, ed. Hoskin, *The First Autobiography*, p. 29.

80 *Caroline Herschel's Autobiographies*, ed. Hoskin, *The Second Autobiography*, p. 114.

81 *Caroline Herschel's Autobiographies*, ed. Hoskin, *The First Autobiography*, p. 46; p. 34.

82 Caroline Herschel to John Herschel, 25 September 1827, BL Eg 3762, Herschel Papers, 71–2.

CHAPTER II

1 *Caroline Herschel's Autobiographies*, ed. Hoskin, *The First Autobiography*, p. 47.

2 *Caroline Herschel's Autobiographies*, ed. Hoskin, *The Second Autobiography*, p. 116.

3 *Caroline Herschel's Autobiographies*, ed. Hoskin, *The First Autobiography*, p. 47.

4 *Caroline Herschel's Autobiographies*, ed. Hoskin, *The First Autobiography*, p. 47.

5 *Caroline Herschel's Autobiographies*, ed. Hoskin, *The First Autobiography*, p. 47; *The Second Autobiography*, p. 116.

6 Caroline Herschel to Margaret Herschel, 24 September 1838, BL Eg 3762, Herschel Papers, 32–3.

7 Cyril Ehrlich, *The Music Profession in Britain Since the Eighteenth Century: A Social History* (Oxford: Clarendon Press, 1985), p. 2.

8   Thomas Bauman, 'Courts and Municipalities in North Germany', in Neal Zaslaw, ed., *The Classical Era: From the 1740s to the end of the eighteenth century* (Basingstoke: Macmillan, 1989), pp. 240–67; p. 241.

9   *Caroline Herschel's Autobiographies*, ed. Hoskin, *The First Autobiography*, p. 44.

10 *Caroline Herschel's Autobiographies*, ed. Hoskin, *The First Autobiography*, p. 47; *The Second Autobiography*, p. 117.

11 *Caroline Herschel's Autobiographies*, ed. Hoskin, *The First Autobiography*, p. 47.

12 *Caroline Herschel's Autobiographies*, ed. Hoskin, *The Second Autobiography*, p. 117.

13 *Caroline Herschel's Autobiographies*, ed. Hoskin, *The First Autobiography*, p. 47.

14 *Caroline Herschel's Autobiographies*, ed. Hoskin, *The First Autobiography*, p. 47.

15 *Caroline Herschel's Autobiographies*, ed. Hoskin, *The Second Autobiography*, p. 117.

16 *Caroline Herschel's Autobiographies*, ed. Hoskin, *The First Autobiography*, p. 48; *The Second Autobiography*, p. 117.

17 Incidents detailed in *The Second Autobiography*, pp. 117–18.

18 *Caroline Herschel's Autobiographies*, ed. Hoskin, *The Second Autobiography*, p. 118. See Sophie von La Roche's account of her visit to Britain and her comments on shopping in London in *Sophie in London* (1788), trans. Clare Williams.

19 See Michael Hoskin's footnote to the text of *The Second Autobiography*, p. 118.

20 For more on eighteenth-century Bath, see John Brewer, *The Pleasures of the Imagination: English Culture in the Eighteenth Century* (London: HarperCollins, 1997) and Roy Porter, *English Society in the Eighteenth Century* (Harmondsworth: Penguin, 1990).

21 Tobias Smollett, *The Expedition of Humphry Clinker* (1771), ed. Lewis M. Knapp (Oxford: Oxford University Press, 1984), pp. 36–7.

22 A. J. Turner, *Science and Music in Eighteenth-Century Bath: An Exhibition in the Holburne of Menstrie Museum, Bath, 22 September 1977 – 29 December 1977* (Bath: University of Bath, 1977), p. 15.

23 The facts about music and musicians in eighteenth-century Bath can be found in Ehrlich, *The Music Profession in Britain Since the Eighteenth Century*, p. 23.

24 Ehrlich, *The Music Profession*, p. 3.

25 Ehrlich, *The Music Profession*, p. 3.

26 Ehrlich, *The Music Profession*, pp. 5–6; p. 18.

27 Ehrlich, *The Music Profession*, p. 17.

28 This letter is quoted in Turner, *Science and Music in Eighteenth-Century Bath*, p. 24.

29 Turner, *Science and Music in Eighteenth-Century Bath*, p. 31

30 Turner, *Science and Music in Eighteenth-Century Bath*, p. 31

31 Turner, *Science and Music in Eighteenth-Century Bath*, p. 31.

32 *Caroline Herschel's Autobiographies*, ed. Hoskin, *The First Autobiography*, p. 49; *The Second Autobiography*, p. 119.

33 *Caroline Herschel's Autobiographies*, ed. Hoskin, *The Second Autobiography*, p. 119.

34 *Caroline Herschel's Autobiographies*, ed. Hoskin, *The First Autobiography*, p. 49.

35 *Caroline Herschel's Autobiographies*, ed. Hoskin, *The Second Autobiography*, p. 120; *The First Autobiography*, p. 50.

36 *Caroline Herschel's Autobiographies*, ed. Hoskin, *The First Autobiography*, p. 50.

37 Bridget Hill, *Women, Work and Sexual Politics in Eighteenth-Century England* (Oxford: Oxford University Press, 1989), p. 292.

38 *Caroline Herschel's Autobiographies*, ed. Hoskin, *The Second Autobiography*, p. 120.

39 [Hannah Woolley], *Accomplish'd Lady's Delight in Preserving, Physik, Beautifying, Cookery and Gardening* (1719), Preface.

40 *Caroline Herschel's Autobiographies*, ed. Hoskin, *The First Autobiography*, p. 50; *The Second Autobiography*, p. 119.

41 *Caroline Herschel's Autobiographies*, ed. Hoskin, *The Second Autobiography*, p. 121.

42 *Caroline Herschel's Autobiographies*, ed. Hoskin, *The First Autobiography*, p. 51.

43   *Caroline Herschel's Autobiographies*, ed. Hoskin, *The Second Autobiography*, p. 121.

44   *Caroline Herschel's Autobiographies*, ed. Hoskin, *The Second Autobiography*, p. 123.

45   *Caroline Herschel's Autobiographies*, ed. Hoskin, *The Second Autobiography*, p. 123.

46   *Caroline Herschel's Autobiographies*, ed. Hoskin, *The Second Autobiography*, pp. 123–4.

47   Frances Burney, *Evelina; or A Young Lady's Entrance into the World* (1778), ed. Margaret Anne Doody (London: Penguin, 2004), p. 92.

48   *Caroline Herschel's Autobiographies*, ed. Hoskin, *The First Autobiography*, p. 53.

49   *Caroline Herschel's Autobiographies*, ed. Hoskin, *The Second Autobiography*, p. 125. Hoskin labels the London trip a 'traumatic experience', but Herschel herself suggests that it was both disturbing and curiously fascinating.

50   *Caroline Herschel's Autobiographies*, ed. Hoskin, *The First Autobiography*, p. 51.

51   *Caroline Herschel's Autobiographies*, ed. Hoskin, *The First Autobiography*, p. 55.

52   *Caroline Herschel's Autobiographies*, ed. Hoskin, *The First Autobiography*, p. 55.

53   Quoted in Margaret Alic, *Hypatia's Heritage: A History of Women in Science from Antiquity to the Late Nineteenth Century* (London: The Women's Press, 1986), p. 93.

54   For further detail, see Alice N. Walters, 'Conversation
      Pieces: Science and Politeness in Eighteenth-Century
      England', *History of Science*, 35 (1997), 121–54; 125.

55   For more discussion of Martin's text, see Patricia
      Fara, *An Entertainment for Angels: Electricity in the
      Enlightenment* (Cambridge: Icon Books, 2002).

56   Aphra Behn, *A Discovery of New Worlds* (1688), in
      The Works of Aphra Behn, ed. Janet Todd (London:
      William Pickering, 1993), Vol. IV, p. 156. On Behn's
      text, see Aileen Douglas, 'Popular Science and the
      Representation of Women: Fontenelle and After',
      Eighteenth-Century Life, 18.2 (1994), 1–14.

57   Elizabeth Carter to Catherine Talbot, Canterbury, 20
      March 1747, in *A Series of Letters between Mrs Elizabeth
      Carter and Miss Catherine Talbot, from the year 1741 to
      1770*, ed. Montagu Pennington, 3 volumes (London:
      F.C. and J. Rivington, 1819), I, p. 128.

58   Mary Somerville, *Preliminary Dissertation to
      Mechanism of the Heavens* (London: John Murray,
      1831), pp. v–lxx; p. vi.

59   *Caroline Herschel's Autobiographies*, ed. Hoskin, *The
      First Autobiography*, p. 55.

60   Robert B. Shoemaker, *Gender in English Society 1650–
      1850: The Emergence of Separate Spheres?* (London
      and New York: Longman, 1998), p. 194.

61   *Caroline Herschel's Autobiographies*, ed. Hoskin, *The
      First Autobiography*, p. 53.

62   *Caroline Herschel's Autobiographies*, ed. Hoskin, *The
      Second Autobiography*, p. 128.

63 *Caroline Herschel's Autobiographies*, ed. Hoskin, *The First Autobiography*, p. 56.

64 *Caroline Herschel's Autobiographies*, ed. Hoskin, *The First Autobiography*, p. 55.

65 *Caroline Herschel's Autobiographies*, ed. Hoskin, *The Second Autobiography*, p. 129.

66 Frank Brown, *Caroline Herschel as a Musician* (Bath: The William Herschel Museum, 2000), p. 9.

67 Anthony Hicks and Gerald Abraham, 'Oratorio and Related Forms', in *New Oxford History of Music, Volume VI: Concert Music 1630–1750*, ed. Gerald Abraham (Oxford: Oxford University Press, 1986), pp. 23–96; p. 86.

68 Brown, *Caroline Herschel as a Musician*, p. 9; Caroline *Herschel's Autobiographies*, ed. Hoskin, *The First Autobiography*, p. 56.

69 Brown, *Caroline Herschel as a Musician*, p. 9.

70 Brown, *Caroline Herschel as a Musician*, p. 13.

71 In *Caroline Herschel as a Musician*, Brown too offers this reasoning behind Herschel's position in the concert. He does not, however, stress the fact that this was a benefit performance for William Herschel, which makes the necessity for a polished performance all the more urgent. In the section 'How good a singer was Caroline?', Brown writes: 'The fact that William had spent considerable time and effort to train her in the rôle of public singer in his concerts showed that he believed her to have considerable potential. He was

unlikely to have risked appearing with her on the same concert platform and thereby compromising his own musical career and livelihood if she were of an inferior standard.' (p. 13)

72 For more on the bitter Herschel–Linley rivalry in the 1770s, see the entry on William Herschel in the *New Grove Dictionary of Music and Musicians*, Volume 8, ed. Stanley Sadie (London: Macmillan, 1980), pp. 520–2; p. 521, and Ehrlich, *The Music Profession in England*, p. 24. Linley's daughter Elizabeth, the famous singer, married the playwright Richard Brinsley Sheridan after an elopement. The Herschel–Linley rivalry, according to Ehrlich, centred around Linley's desire to monopolise benefit concerts in Bath with his daughter as an exclusive attraction (p. 24).

73 For the contemporary popularity of the *Messiah*, see Anthony Hicks and Gerald Abraham, 'Oratorio and Related Forms', in *New Oxford History of Music*, p. 84.

74 *Caroline Herschel's Autobiographies*, ed. Hoskin, *The First Autobiography*, p. 57.

75 Alice Corkran, *The Romance of Women's Influence* (London: Blackie and Son Limited, 1906), p. 169.

76 Michael Hoskin, *The Herschel Partnership as viewed by Caroline* (Cambridge: Science History Publications Limited, 2003), p. 42; *Caroline Herschel's Autobiographies*, ed. Hoskin, *The First Autobiography*, p. 57.

77 Brown, *Caroline Herschel as a Musician*, p. 12.

78   *Caroline Herschel's Autobiographies*, ed. Hoskin, *The Second Autobiography*, p. 129.
79   Mrs John Herschel, *Memoir and Correspondence of Caroline Herschel* (London: John Murray, 1876), p. 30.
80   Brown, *Caroline Herschel as a Musician*, pp. 9–10.

CHAPTER III

1   *Caroline Herschel's Autobiographies*, ed. Hoskin, *The First Autobiography*, p. 59.
2   19 New King Street, Bath is now the home of The William Herschel Museum.
3   *Caroline Herschel's Autobiographies*, ed. Hoskin, *The First Autobiography*, p. 60.
4   *Caroline Herschel's Autobiographies*, ed. Hoskin, *The First Autobiography*, p. 61.
5   Hoskin's note 43 to *The First Autobiography*, p. 61.
6   *Caroline Herschel's Autobiographies*, ed. Hoskin, *The First Autobiography*, p. 60; p. 61.
7   *Caroline Herschel's Autobiographies*, ed. Hoskin, *The First Autobiography*, p. 61.
8   Bridget Hill, *Women Alone: Spinsters in England, 1660–1850* (New Haven and London: Yale University Press, 2001), p. 40.
9   Bridget Hill, *Women, Work and Sexual Politics in Eighteenth-Century England* (Oxford: Oxford University Press, 1989), p. 85.
10   *Caroline Herschel's Autobiographies*, ed. Hoskin, *The First Autobiography*, p. 61.

11 Sir Patrick Moore, *Caroline Herschel: Reflected Glory* (Bath: The William Herschel Museum, 1988), p. 10.

12 See Hoskin's note 10 to *Caroline Herschel's Autobiographies, The First Autobiography*, p. 52.

13 For the worries over the fate of the optical industry in the late eighteenth century, particularly in London, see John North, *The Fontana History of Astronomy and Cosmology* (London: Fontana Press, 1994), p. 387.

14 *Caroline Herschel's Autobiographies*, ed. Hoskin, *The First Autobiography*, pp. 62–3. Naming the new planet Georgium Sidus ('George's Star') after the King and therefore exciting the interest of a potential patron was a canny move on William Herschel's part.

15 See n. 48, *Caroline Herschel's Autobiographies*, ed. Hoskin, *The First Autobiography*, p. 63.

16 Henry C. King, *The History of the Telescope* (London: Charles Griffin and Company, 1955), p. 126.

17 Letter from William to Caroline Herschel, 3 June 1782; Mrs John Herschel, *Memoir and Correspondence of Caroline Herschel* (London: John Murray, 1876), p. 47.

18 *Caroline Herschel's Autobiographies*, ed. Hoskin, *The First Autobiography*, p. 62.

19 *Caroline Herschel's Autobiographies*, ed. Hoskin, *The First Autobiography*, p. 63.

20 Cyril Ehrlich, *The Music Profession in Britain Since the Eighteenth Century*, p. 24.

21 Frank Brown, *Caroline Herschel as a Musician* (Bath: The William Herschel Museum, 2000), p. 10.

22 Ehrlich, *The Music Profession*, p. 24.

23 Ehrlich, *The Music Profession*, p. 24.

24 For Herschel's income from music, see his entry in the *New Grove Dictionary of Music and Musicians*, Volume 8, ed. Stanley Sadie (London: Macmillan, 1980), pp. 520–2; p. 522; and Michael Hoskin, 'Vocations in Conflict: William Herschel in Bath, 1766–1782', *History of Science*, 41.3 (2003), 315–33; 333.

25 See Charles Babbage's lamentation at the state of science in his polemic *Reflection on the Decline of Science in England and on Some of its Causes* (London: B. Fellowes and J. Booth, 1830). See also James A. Secord's *Victorian Sensation: The Extraordinary Publication, Reception, and Secret Authorship of Vestiges of the Natural History of Creation* (Chicago and London: University of Chicago Press, 2000) on the limits of the term 'professional science' as opposed to 'commercial science' at this period, especially pp. 437–8.

26 Caroline Herschel to John Herschel, April 1827, BL Eg 3761, Herschel Papers, 60–1.

27 *Caroline Herschel's Autobiographies*, ed. Hoskin, *The First Autobiography*, p. 66.

28 *Caroline Herschel's Autobiographies*, ed. Hoskin, *The First Autobiography*, p. 68.

29 Ehrlich, *The Music Profession*, p. 31.

30 *Caroline Herschel's Autobiographies*, ed. Hoskin, *The First Autobiography*, p. 61.

31   *Caroline Herschel's Autobiographies*, ed. Hoskin, *The First Autobiography*, p. 68.

32   *Court and Private Life in the Time of Queen Charlotte: Being the Journals of Mrs Papendiek, Assistant Keeper of the Wardrobe and Reader to her Majesty*, ed. Mrs Vernon Delves Broughton, 2 vols (London: Richard Bentley & Son, 1887), Volume I, pp. 247–8.

33   *Lady Sarah Pennington, An Unfortunate Mother's Advice to her Daughters (1761), in The Young Lady's Pocket Library, or Parental Monitor; Containing I. Dr Gregory's Father's Legacy to his Daughters. II. Lady Pennington's Unfortunate Mother's Advice to Her Daughters. III. Marchioness de Lambert's Advice of a Mother to Her Daughter. IV. Moore's Fables for the Female Sex* (Dublin: J. Archer, 1790), ed. Vivien Jones (Bristol: Thoemmes Press, 1995).

34   *Caroline Herschel's Autobiographies*, ed. Hoskin, *The Second Autobiography*, p. 119.

35   See Michael Hoskin, *The Herschel Partnership as viewed by Caroline* (Cambridge: Science History Publications Limited, 2003), p. 39. I do not agree with Hoskin that Marsh must have necessarily been incorrect in his remembrance of the events. He may have become confused about the telescopes, but the chance of a layman retaining the details of a conversation over dinner about the operations of scientific instruments would be very slight. Why, however, would he not remember correctly the suggestion that astronomy

was a family enterprise in the Herschel household? William may have been trying to give him the impression that it was a joint effort, or that observing, as he noted himself, was carried out by Caroline too.

36 *Caroline Herschel's Autobiographies*, ed. Hoskin, *The First Autobiography*, p. 71.

37 Caroline Herschel to Margaret Herschel, 2 June 1842, BL Eg 3762, Herschel Papers, 102–03.

38 Mrs John Herschel, *Memoir and Correspondence of Caroline Herschel* (London: John Murray, 1876), p. 71.

39 Mrs John Herschel, *Memoir and Correspondence of Caroline Herschel*, pp. 143–4.

40 *Caroline Herschel's Autobiographies*, ed. Hoskin, *The First Autobiography*, p. 71.

41 *Caroline Herschel's Autobiographies*, ed. Hoskin, *The First Autobiography*, p. 71.

42 For more on the Herschels and the pursuit of nebulae, see North, *The Fontana History of Astronomy and Cosmology*, p. 403.

43 King, *The History of the Telescope*, p. 127.

44 Hoskin, *The Herschel Partnership*, p. 63.

45 MS page from Herschel's notes on her sweeps of 26 February 1783, reproduced in *Caroline Herschel's Autobiographies*, ed. Hoskin, *The First Autobiography*, p. 72.

46 Margaret Bullard, 'My Small Newtonian Sweeper – Where is it Now?', *Notes & Records of the Royal Society of London*, 42.2 (1988), 139–48; 143.

47 Nevil Maskelyne to Edward Pigott, 6 December 1793, quoted, without the date, in *Caroline Herschel's Autobiographies*, ed. Hoskin, *The First Autobiography*, p. 70, and with the date identified in Bullard, 'My Small Newtonian Sweeper', 144. Hoskin notes further that Herschel's telescope would end up on her nephew and William's son John Herschel's trip to the Cape of Good Hope from 1834–8, where he used it to familiarise himself with the skies of the southern hemisphere.

48 *Caroline Herschel's Autobiographies*, ed. Hoskin, *The First Autobiography*, p. 76; n. 68, p. 73.

49 *Caroline Herschel's Autobiographies*, ed. Hoskin, *The First Autobiography*, p. 76.

50 *Caroline Herschel's Autobiographies*, ed. Hoskin, *The First Autobiography*, p. 78.

51 *Caroline Herschel's Autobiographies*, ed. Hoskin, *The First Autobiography*, p. 73.

52 *Caroline Herschel's Autobiographies*, ed. Hoskin, *The First Autobiography*, p. 76.

53 *Caroline Herschel's Autobiographies*, ed. Hoskin, *The First Autobiography*, pp. 76–7.

54 *Caroline Herschel's Autobiographies*, ed. Hoskin, *The First Autobiography*, p. 77.

55 *Caroline Herschel's Autobiographies*, ed. Hoskin, *The First Autobiography*, p. 78.

56 *Caroline Herschel's Autobiographies*, ed. Hoskin, *The First Autobiography*, pp. 79–81.

57   *Caroline Herschel's Autobiographies*, ed. Hoskin, *The First Autobiography*, p. 82.

58   *Caroline Herschel's Autobiographies*, ed. Hoskin, *The First Autobiography*, p. 83.

59   *Caroline Herschel's Autobiographies*, ed. Hoskin, *The First Autobiography*, p. 86.

60   *Court and Private Life in the Time of Queen Charlotte: Being the Journals of Mrs Papendiek*, Volume I, p. 251.

61   *Caroline Herschel's Autobiographies*, ed. Hoskin, *The First Autobiography*, p. 86.

62   Caroline Herschel to Margaret Herschel, letter post-marked 5 June 1843, BL Eg 3762, Herschel Papers, 128–9.

63   Caroline Herschel to Alexander Aubert, 2 August 1786, in Mrs John Herschel, Memoir, p. 67.

64   Caroline Herschel to Dr Charles Blagden, 2 August 1786, in Mrs John Herschel, Memoir, p. 65.

65   Agnes M. Clerke, *The Herschels and Modern Astronomy* (London: Cassell and Company, 1895), p. 140.

66   *Caroline Herschel's Autobiographies*, ed. Hoskin, *The First Autobiography*, p. 90.

67   Adam Smith, *The Principles which Lead and Direct Philosophical Enquiries; Illustrated by the History of Astronomy*, in *Essays on Philosophical Subjects* (London and Edinburgh: T. Cadell/W. Davies and W. Creech, 1795), ed. W.P.D. Wightman and J.C. Bryce (Indianapolis, IN: Liberty Fund, 1982), pp. 33–105; p. 102.

68 David W. Hughes, 'Caroline Lucretia Herschel – comet huntress', *Journal of the British Astronomical Association*, 109.2 (1999), 78–85; 78.

69 Letter from Nevil Maskelyne to William Herschel, 1782, quoted in Roberta J.M. Olson and Jay M. Pasachoff, *Fire in the Sky: Comets and Meteors, the Decisive Centuries in British Art and Science* (Cambridge: Cambridge University Press, 1998), p. 101.

70 Alexander Aubert to Caroline Herschel, 7 August 1786, in Mrs John Herschel, *Memoir*, p. 69.

71 For more on the historical development of the meaning of comets, see Sara J. Schechner, *Comets, Popular Culture, and the Birth of Modern Cosmology* (Princeton, NJ: Princeton University Press, 1997). Quotation from Schechner, p. 214.

72 See Harriet Guest, *Small Change: Women, Learning, Patriotism, 1750–1810* (Chicago and London: University of Chicago Press, 2000), especially 'Part Three: Femininity and National Feeling in the 1770s and 1780s', pp. 155–271.

73 'An Account of Mrs Anna-Laetitia Barbauld', *European Magazine, and London Review*, IX (March 1786), 139–40; 139.

74 Michael Hoskin, 'Caroline Herschel: Assistant Astronomer or Astronomical Assistant?', *History of Science*, 40.4 (2002), 425–44; 435.

75 *Sophie in London 1786: being the Diary of Sophie v. la*

Roche, trans. Clare Williams (London: Jonathan Cape, 1933), 14 September.

76   *Sophie in London*, p. 192.

77   Maria Edgeworth to Sophy Ruxton, Château de Coppet, 28th September 1820, in *Maria Edgeworth in France and Switzerland: Selections from the Edgeworth Family Letters*, ed. Christina Colvin (Oxford: Clarendon Press, 1979), p. 253.

78   *Sophie in London*, pp. 192–3.

79   *Caroline Herschel's Autobiographies*, ed. Hoskin, *The First Autobiography*, p. 94.

80   See Hill, *Women Alone: Spinsters in England 1660–1850*, pp. 63–4.

81   Hill, *Women, Work and Sexual Politics*, p. 133.

82   Letter from Caroline Herschel to Mary Herschel, 14 October 1824, in Mrs John Herschel, *Memoir*, p. 178.

83   Letter from William Herschel to George III, reproduced in Hoskin, *The Herschel Partnership*, p. 88.

CHAPTER IV

1   *Caroline Herschel's Autobiographies*, ed. Hoskin, *The First Autobiography*, p. 96.

2   Caroline Herschel to Alexander Aubert, 18 April 1790, in Mrs John Herschel, *Memoir and Correspondence of Caroline Herschel* (London: John Murray, 1876), p. 86.

3   Caroline Herschel to Sir Joseph Banks, 17 August 1797, Mrs John Herschel, *Memoir*, pp. 94–5.

4   Sir Harry Englefield to William Herschel, 25 December 1788, Mrs John Herschel, *Memoir*, p. 83.

5   Caroline Herschel to Jérôme de Lalande, 2 July 1790, BL Add 37203, Babbage Papers on Astronomy, 4.

6   Charles Hutton, *A Companion or Supplement to the Ladies Diary, for the year 1791* (London, 1790), 144.

7   *New Annual Register for the year 1793* (London, 1794), 42; 'History of the Sciences for 1797', *The Scientific Magazine, and Freemason's Repository for 1798*, p. 150.

8   *An Historical Miscellany of the Curiosities and Rarities in Nature and Art comprising new and entertaining descriptions of the most surprising volcanoes, caverns, cataracts, whirlpools, waterfalls, earthquakes, thunder, lightning and other wondrous and stupendous phenomena of nature. Forming a rich and comprehensive view of all that is interesting and curious in every part of the habitable world* (London: By the Proprietor, 1794–1800), Volume IV, p. 14; John Payne, *Geographical Extracts, forming a general view of the earth and nature* (London: G.G. and J. Robinson, 1796), p. 59.

9   Entry on 'Dr Herschel', *British Public Characters of 1798* (London: R. Phillips, 1798), p. 366.

10 Caroline Herschel to Nevil Maskelyne, 3 March 1794, BL Add 37203, Babbage Papers on Astronomy, 6.

11 *Court and Private Life in the Time of Queen Charlotte: Being the Journals of Mrs Papendiek, Assistant Keeper of the Wardrobe and Reader to her Majesty*, ed. Mrs Vernon Delves Broughton, 2 Volumes (London: Richard Bentley & Son, 1887), Volume I, pp. 275–6.

12 *Court and Private Life in the Time of Queen Charlotte*, p. 282.

13 Memorandum dated 31 December 1798, in Mrs John Herschel, *Memoir*, pp. 98–9.

14 Memorandum of 28 March 1800, in Mrs John Herschel, *Memoir*, p. 107.

15 Caroline Herschel to Margaret Herschel, 29 February 1840, BL Eg 3762, Herschel Papers, 53–4.

16 Journal letter from Frances Burney to Susanna Phillips, 17 September 1787, in *Frances Burney: Journals and Letters*, eds Peter Sabor and Lars E. Troide (Harmondsworth: Penguin, 2001), p. 252.

17 Barthélemi Faujas de St-Fond, *Travels in England and Scotland and the Hebrides; undertaken for the purposes of examining the state of the arts and the sciences, natural history and manners in Great Britain* (London, 1799), 2 volumes, Volume I, p. 65, p. 78.

18 *Travels in England and Scotland*, p. 78.

19 *Travels in England and Scotland*, p. 65.

20 The book is now preserved in The William Herschel Museum, Bath.

21 Caroline Herschel to Margaret Herschel, 6 September 1843, BL Eg 3762, Herschel Papers, 131–2. Herschel suggests that the Princesse, whom scandalmongering contemporaries alleged was a lesbian lover of Marie Antoinette, visited Slough in 1787, but, of course, this mention of her guillotining two weeks after is historically inaccurate. In *The Great Nation: France from Louis XIV to Napoleon* (London: Penguin, 2003),

Colin Jones notes that she was a victim of the culling of around 1,100 to 1,500 prisoners in September 1792 (p. 461).

22 Henry C. King, *The History of the Telescope* (London: Charles Griffin and Company, 1955), p. 130.

23 Caroline Herschel to Margaret Herschel, 9–10 Jan 1840, BL Eg 3762, Herschel Papers, 50–1.

24 Caroline Herschel to Margaret Herschel, 9–10 Jan 1840, BL Eg 3762, Herschel Papers, 50–1.

25 William Herschel, *Description of a Forty Feet Reflecting Telescope, Philosophical Transactions* (London, 1795), reprinted in *The Scientific Papers of William Herschel*, ed. J.L.E. Dreyer (London: Royal Society/Royal Astronomical Society, 1912), p. 486.

26 Dreyer, 'Introduction' to *The Scientific Papers of William Herschel*, xlvii.

27 This account is indebted to King, *The History of the Telescope*, p. 129, and to Herschel's own explanations in *Description of a Forty Feet Reflecting Telescope*.

28 Michael Hoskin, 'Caroline Herschel: Assistant Astronomer or Astronomical Assistant?', *History of Science*, 40.4 (2002), 425–44, 438.

29 *Court and Private Life in the Time of Queen Charlotte*, p. 252.

30 *Sophie in London*, trans. Clare Williams, p. 191.

31 Journal letter from Frances Burney to Susanna Phillips, 17 September 1787, in *Frances Burney: Journals and Letters*, p. 252.

32   Michael Hoskin, *The Herschel Partnership as viewed by Caroline* (Cambridge: Science History Publications Limited, 2003), pp. 109–110.

33   See, for example, Marina Benjamin, 'Elbow Room: Women Writers on Science, 1790–1840', in *Science and Sensibility: Gender and Scientific Enquiry 1780–1945*, ed. Marina Benjamin (Oxford: Basil Blackwell, 1991), pp. 27–59.

34   Marina Benjamin, 'A Question of Identity', in *A Question of Identity: Women, Science, and Literature* (New Brunswick, NJ: Rutgers University Press, 1993), ed. Marina Benjamin, pp. 1–21; p. 8.

35   Kathryn A. Neeley, 'Women as Mediatrix: Women as Writers on Science and Technology in the Eighteenth and Nineteenth Centuries', *IEEE Transactions on Professional Communication*, 35.4 (December 1992), 208–16.

36   Adam Smith, *The Theory of Moral Sentiments* (1759), eds D.D. Raphael and A.L. Macfie (Indianapolis, IN: Liberty Fund, 1984), pp. 123–5; p. 125.

37   William Herschel, *Description of a Forty Feet Reflecting Telescope*, p. 52.

38   William Herschel, *A third catalogue of the comparative brightness of the stars; with an introductory account of an index to Mr Flamsteed's observations of the fixed stars, Philosophical Transactions* (London, 1797), p. 7.

39   Marilyn Bailey Ogilvie, 'Caroline Herschel's Contributions to Astronomy', *Annals of Science*, 32 (1975), 149–61; 156.

40   Hoskin, *The Herschel Partnership*, p. 102.

41   Hoskin, *The Herschel Partnership*, pp. 102–3.

42   Carolina Herschel, *Catalogue of Stars, taken from Mr Flamsteed's Observations contained in the second volume of the Historia Coelestis, and not inserted in the British Catalogue, with an Index to point out every observation in that volume belonging to the stars of the British Catalogue. To which is added, a collection of errata that should be noticed in the same volume* (London: Royal Society, 1798) and Hoskin, *The Herschel Partnership*, p. 102.

43   Edward Pigott, Esq. to Caroline Herschel, 30 April 1799, Mrs John Herschel, *Memoir*, p. 103.

44   Ogilvie, 'Caroline Herschel's Contributions to Astronomy', 157.

45   Hoskin, *The Herschel Partnership*, p. 103.

46   Caroline Herschel to Nevil Maskelyne, September 1798, Mrs John Herschel, *Memoir*, p. 96.

47   Hoskin, *The Herschel Partnership*, p. 103.

48   J.L.E. Dreyer, quoted in Constance A. Lubbock, *The Herschel Chronicle: The Life Story of William Herschel and his Sister Caroline Herschel* (Cambridge: Cambridge University Press, 1933), p. 160.

49   Mrs John Herschel, *Memoir*, pp. 95–6.

50   Sir Patrick Moore, *Caroline Herschel: Reflected Glory* (Bath: The William Herschel Museum, 1988), p. 16.

51   Mrs John Herschel, *Memoir*, p. 117.

52   Hoskin, *The Herschel Partnership*, p. 119.

53 Agnes M. Clerke, *The Herschels and Modern Astronomy* (London: Cassell and Company, 1895), p. 126; p. 127.

54 Meetings described in entries for 12 May 1813 and 4 December 1814, Mrs John Herschel, *Memoir*, pp. 121–2.

55 Invitation received on 14 July, party held on 17 July 1814, Mrs John Herschel, *Memoir*, p. 126.

56 *Ladies' Astronomy*, trans. from the French of Jérôme de Lalande by Mrs W. Pengree (London: Darton, Harvey, and Darton, 1815), x–xi.

57 See the very important article by Paula Findlen, 'Science as a Career in Enlightenment Italy: The Strategies of Laura Bassi', *Isis*, 84 (1993), 441–69. Also, for European context, see Londa Schiebinger, *The Mind Has No Sex?: Women in the Origins of Modern Science* (Cambridge, MA and London: Harvard University Press, 1989).

58 *Ladies' Astronomy*, x.

59 Hoskin, *The Herschel Partnership*, p. 127.

60 The King of Hanover, the former Duke of Cumberland, thought John was Caroline's son (Caroline Herschel to John Herschel, 11 June 1837, BL Eg 3762, Herschel Papers, 20–1). Caroline Herschel to John Herschel, 3 February 1829, BL Eg 3761, Herschel Papers, 97–8.

61 Day-Book, October 1821, Mrs John Herschel, *Memoir and Correspondence*, pp. 132–3.

62 Memorandum for 19 December 1808, Mrs John Herschel, *Memoir*, p. 116.

63 Caroline Herschel to Mary Herschel, 14 October 1824, Mrs John Herschel, *Memoir*, p. 178; Hoskin, *The Herschel Partnership*, p. 120.

64 Mrs John Herschel, *Memoir*, p. 139; Clerke, *The Herschels and Modern Astronomy*, p. 130.

65 Obituary of William Herschel, *Gentleman's Magazine*, 92 (September 1822), 274–6; 275.

66 'Additions to the Obituary', *Gentleman's Magazine*, 92 (December 1822), 650.

67 Caroline Herschel to Margaret Herschel; date, 29 February 1840, written at the top of the page in a different pen, BL Eg 3762, Herschel Papers, 53–4.

68 Patricia Fara, *Pandora's Breeches*, p. 163; p. 151.

69 Caroline Herschel to Margaret Herschel, 11 August 1831, BL Eg 3761, 141–2.

70 Caroline Herschel to John Herschel, 24 June 1823, BL Eg 3761, Herschel Papers, 16–17.

71 John Herschel to Margaret Herschel, 19 June 1832, Mrs John Herschel, *Memoir and Correspondence*, p. 255. Also Günther Buttmann, *The Shadow of the Telescope: A Biography of John Herschel*, trans. E. J. Pagel (Guildford and London: Lutterworth Press, 1974), p. 73.

72 Caroline Herschel to John Herschel, 26 December 1822, BL Eg 3761, Herschel Papers, 11–12.

73 Hoskin, 'Caroline Herschel: Assistant Astronomer or Astronomical Assistant?', 425.

74 Hoskin, *The Herschel Partnership*, p. 135.

75 Caroline Herschel to John Herschel, 11 August 1823, BL Eg 3761, Herschel Papers, 19–20.

76 Caroline Herschel to Miss Baldwin, 23 August 1824, BL Eg 3761, Herschel Papers, 22–3.

77 Draft of a letter from John Herschel to Caroline Herschel, 6 December 1824, BL Eg 3761, Herschel Papers, 30.

78 Caroline Herschel to John Herschel, 14 January 1825, BL Eg 3761, Herschel Papers, 31–2.

79 Caroline Herschel to John Herschel, 7 March 1825, BL Eg 3761, Herschel Papers, 33–4.

80 Caroline Herschel to John Herschel, 27 March 1825, BL Eg 3761, Herschel Papers, 37.

81 Moore, *Caroline Herschel: Reflected Glory*, p. 17.

82 Caroline Herschel to John Herschel, 1 February 1826, BL Eg 3761, Herschel Papers, 44–5.

83 Caroline Herschel to Margaret Herschel, 4 December 1832, BL Eg 3761, Herschel Papers, 175–6.

84 Caroline Herschel to Margaret Herschel, 6 September 1833, BL Eg 3761, Herschel Papers, 189–90.

85 Caroline Herschel to John Herschel, 1 May 1834, BL Eg 3762, Herschel Papers, 1–2.

86 Caroline Herschel to John Herschel, 11 (7 crossed out) September 1834, BL Eg 3762, Herschel Papers, 4–5.

87 John Herschel to Caroline Herschel, 6 June 1834 and 11 September 1834, Mrs John Herschel, *Memoir and*

Correspondence, p. 258; p. 269. Hoskin suggests that Caroline wanted William's diagnosis of celestial holes confirmed, but, when the letter is quoted in full, she also draws attention to this problem as something she afterwards considered herself. The desire to be right is both out of loyalty to William and the need to uphold her own opinions (*The Herschel Partnership*, p. 143).

88 Caroline Herschel to John Herschel, 20 April 1832, BL Eg 3761, Herschel Papers, 158–9.

89 Caroline Herschel to John Herschel, 3 April 1828, BL Eg 3761, Herschel Papers, 78–9. For the explosion of scientific societies in this period, see Ian Inkster and Jack Morrell, eds, *Metropolis and Province: Science in British Culture, 1780–1850* (London: Hutchinson and Co., 1983).

90 John Herschel to Caroline Herschel, 18 April 1825, Mrs John Herschel, *Memoir*, p. 188.

91 'Address to the AS, by J. South, Esq., on presenting the Honorary Medal to Miss C. Herschel, at its 8th General Meeting, Feb 8th, 1828', reproduced in Mrs John Herschel, *Memoir and Correspondence*, pp. 222–6.

92 *The Collected Works of Mary Somerville*, in 9 volumes, have recently been published, edited by James A. Secord (Bristol: Thoemmes Press, 2004). Secord has, importantly, collected all Somerville's original scientific papers and reviews in the first volume of *The Collected Works*. These include early responses to

prize questions in the *New Series of the Mathematical Repository*, a letter to Josephine Butler on 'The Teaching of Science', and the experiments detailed in the *Philosophical Transactions* and the *Comptes rendus hebdomadaires des séances de l'Academie des Sciences.*

93 William Whewell, 'Review of *On the Connexion of the Physical Sciences* by Mrs Somerville', *Quarterly Review*, LI (1834), 54–68. Reiterating an idea he had first presented at the previous year's British Association for the Advancement of Science meeting, Whewell proposed a comprehensive title for those engaged in all forms of scientific study at a time when the sciences were beginning to splinter from the all-inclusive natural philosophy into more professional fields of physics, chemistry, natural history and so on. Paradoxically, while Whewell valued the elements of 'connexion' between what were soon to become different disciplines, and thus harked back to the glory of interdisciplinary unity, the title he chose became one which would be universally adopted as the movement towards professionalisation grew. 'Scientist', suggested Whewell, could be employed in similar fashion to artist or economist.

94 Caroline Herschel to Margaret Herschel, 4 September 1844, BL Eg 3762 Herschel Papers, 154–5.

95 Caroline Herschel to Margaret Herschel, 20 April 1832, BL Eg 3761, Herschel Papers 158–9; 7 September 1832, 171–2; 4 December 1832, 175–6.

96   Caroline Herschel to John Herschel, 11 June 1837, BL Eg 3762, Herschel Papers, 20–1.

97   Caroline Herschel to Augustus de Morgan, Secretary of the Royal Astronomical Society, 9 March 1835, BL Eg 3762, Herschel Papers, 6.

98   *Queen of Science: Personal Recollections of Mary Somerville*, ed. Dorothy McMillan (Edinburgh: Canongate, 2001), p. 86; p. 141.

99   'Extract from the Report of the Council of the Astronomical Society to the Annual Meeting, February 13 1835', Mrs John Herschel, *Memoir*, pp. 226–7.

100 William Rowan Hamilton to Caroline Herschel, 4 December 1838, annotated by Herschel on 7 January 1839, BL Eg 3762, Herschel Papers, 37.

101 Caroline Herschel to Margaret Herschel, 1 October 1846, BL Eg 3762, Herschel Papers, 199. This is Caroline's last extant letter. The letter from Humboldt on 25 September 1846 announced the award of the medal; Mrs John Herschel, *Memoir*, p. 336.

102 John Herschel to Caroline Herschel, 28 May 1828, Mrs John Herschel, *Memoir*, p. 227.

103 Fara, *Pandora's Breeches*, p. 163.

104 Herschel refers to herself and to Alexander as tools in a letter to John Herschel, 14 January 1825, BL Eg 3761, Herschel Papers, 31–2.

105 Caroline Herschel to John Herschel, 21 August 1828, BL Eg 3761, Herschel Papers, 89–90.

106 Caroline Herschel to John Herschel, 3 June 1828, BL Eg 3761, Herschel Papers, 82–3.

107 Caroline Herschel to John Herschel, 23 June 1828, BL Eg 3761, Herschel Papers, 84–5.

108 Caroline Herschel to Margaret Herschel, 29 April 1835, Herschel Papers, BL Eg 3762, 7–8; Caroline Herschel to John Herschel, 17 December 1838, BL Eg 3762, Herschel Papers, 36.

109 William Rowan Hamilton to Caroline Herschel, 4 December 1838, annotated by Herschel on 7 January 1839, BL Eg 3762, Herschel Papers, 37.

110 Caroline Herschel to Margaret Herschel, 7 January 1839, BL Eg 3762, Herschel Papers, 60–1.

111 Caroline Herschel to Margaret Herschel, 29 April 1835, BL Eg 3762, Herschel Papers, 7–8.

112 Caroline Herschel to Margaret Herschel, 30 April 1840, BL Eg 3762, Herschel Papers, 57–8.

113 Caroline Herschel to Margaret Herschel, 3 July 1844, BL Eg 3762, Herschel Papers, 150–1.

114 Caroline Herschel to John Herschel, 3 May 1825, BL Eg 3761, Herschel Papers, 39–40.

115 Caroline Herschel to Margaret Herschel, 7 July 1842, BL Eg 3762, Herschel Papers, 104–5. Herschel had, of course, discovered eight comets, for five of which she has priority.

116 Caroline Herschel to John Herschel, 14 January 1825, BL Eg 3761, Herschel Papers, 31–2; structured as in the original.

117 'Obituary – Miss Herschel', from the *Athenaeum* and reprinted in the *Gentleman's Magazine* (April 1848), 442–3; 442.

118 Tombstone epigraph reprinted in the original German in Mrs John Herschel, *Memoir*, p. 351.

119 Caroline Herschel to Margaret Herschel, 7 July 1842, BL Eg 3762, Herschel Papers, 104–5. In *The Herschels and Modern Astronomy*, Agnes Clerke shortens and tidies up Herschel's letter, thus losing the sense (p. 139).

CONCLUSION

1   For the Royal Institution, see Morris Berman, *Social Change and Scientific Organisation: The Royal Institution, 1799–1844* (London: Heinemann, 1978).

2   Joanna Baillie to Mary Somerville, 1 February 1832, in *Queen of Science: Personal Recollections of Mary Somerville*, ed. Dorothy McMillan (Edinburgh: Canongate, 2001), p. 165.

3   Sir Joseph Banks to Caroline Herschel, 20 April 1790, in Mrs John Herschel, *Memoir*, p. 87.

4   Emma Wallington, 'The physical and intellectual capacities of woman equal to those of men', *Anthropologia*, 1 (1874), 552–65, in Barbara T. Gates, ed., *In Nature's Name: An Anthology of Women's Writing and Illustration, 1780–1930* (Chicago and London: University of Chicago Press, 2002), pp. 32–47; p. 37.

Bibliography

PRIMARY SOURCES

Anon., *Female Rights vindicated; or the Equality of the Sexes Morally and Physically proved* (London: G. Burnet, 1758)

Mary Astell, *The Christian Religion as Profess'd by a Daughter of the Church of England in a Letter to the Right Honourable T.L., C.I.* (London: S.H. for R. Wilkin, 1705)

Jane Austen, *Emma* (1816), ed. Ronald Blythe (Harmondsworth: Penguin, 1985)

Jane Austen, *Sense and Sensibility* (1811), ed. Tony Tanner (Harmondsworth: Penguin, 1986)

British Library Add MS 37203, Babbage Papers on Astronomy

Reflection on the Decline of Science in England and on Some of its Causes (London: B. Fellowes and J. Booth, 1830)

The Works of Aphra Behn, ed. Janet Todd (London: William Pickering, 1993)

British Public Characters of 1798 (London: R. Phillips, 1798)

Frances Burney: Journals and Letters, ed. Peter Sabor and Lars E. Troide (Harmondsworth: Penguin, 2001)

A Series of Letters between Mrs Elizabeth Carter and Miss Catherine Talbot, from the year 1741 to 1770. To which are added, Letters from Mrs Carter to Mrs Vesey, between the years 1763 and 1787; published from the original manuscripts, ed. Montagu Pennington, 3 volumes (London: F.C. and J. Rivington, 1819)

Maria Edgeworth in France and Switzerland: Selections from the Edgeworth Family Letters, ed. Christina Colvin (Oxford: Clarendon Press, 1979)

Barthélemi Faujas de St-Fond, *Travels in England and Scotland and the Hebrides; undertaken for the purpose of examining the state of the arts and the sciences, natural history and manners in Great Britain*, 2 volumes (London, 1799)

Caroline Herschel's Autobiographies, ed. Michael Hoskin (Cambridge: Science History Publications Limited, 2003)

Letters of Caroline Herschel, British Library Egerton MS 3761–3762, Herschel Papers 1822–1866, Volumes I and II

Carolina Herschel, *Catalogue of Stars taken from Mr Flamsteed's Observations contained in the second volume of the Historia Coelestis, and not inserted in the British Catalogue, with an Index to point out every observation in that volume belonging to the stars of the British Catalogue. To which is added, a collection of errata that should be noticed in the same volume* (London: Royal Society, 1798)

Mrs John Herschel, *Memoir and Correspondence of Caroline Herschel* (London: John Murray, 1876)

The Scientific Papers of William Herschel, ed. J.L.E. Dreyer (London: Royal Society/Royal Astronomical Society, 1912)

William Herschel, *A third catalogue of the comparative brightness of the stars; with an introductory account of an index to Mr Flamsteed's observations of the fixed stars, Philosophical Transactions* (London, 1797)

William Herschel, *Description of a Forty Feet Reflecting Telescope, Philosophical Transactions* (London, 1795)

An Historical Miscellany of the Curiosities and Rarities in Nature and Art comprising new and entertaining descriptions of the most surprising volcanoes, caverns, cataracts, whirlpools, waterfalls, earthquakes, thunder, lightning and other wondrous and stupendous phenomena of nature. Forming a rich and comprehensive view of all that is interesting and curious in every part of the habitable world (London: By the Proprietor, 1794–1800)

Ladies' Astronomy, translated from the French of Jerome de Lalande by Mrs W. Pengree (London: Darton, Harvey, and Darton, 1815)

Nicolas Malebranche, *Recherche de la vérité* (1674) (Paris: Galerie de la Sorbonne, 1991)

Court and Private Life in the Time of Queen Charlotte: Being the Journals of Mrs Papendiek, Assistant Keeper of the Wardrobe and Reader to her Majesty, ed. Mrs Vernon Delves Broughton, 2 volumes (London: Richard Bentley and Son, 1887)

John Payne, *Geographical Extracts, forming a general view of the earth and nature* (London: G. G. and J. Robinson, 1796)

Lady Sarah Pennington, *An Unfortunate Mother's Advice to her Daughters* (1761), in *The Young Lady's Pocket Library, or Parental Monitor; Containing I. Dr Gregory's Father's Legacy to his Daughters. II. Lady Pennington's Unfortunate Mother's Advice to Her Daughters. III. Marchioness de Lambert's Advice of a Mother to Her Daughter. IV. Moore's Fables for the Female Sex* (Dublin: J. Archer, 1790), ed. Vivien Jones (Bristol: Thoemmes Press, 1995)

Sophie in London 1786: being the Diary of Sophie v. la Roche, trans Clare Williams (London: Jonathan Cape, 1933)

Jean-Jacques Rousseau, *Emile, Or On Education* (1762), ed. Allan Bloom (Harmondsworth: Penguin, 1991)

Adam Smith, *The Principles which Lead and Direct Philosophical Enquiries; Illustrated by the History of Astronomy, in Essays on Philosophical Subjects* (London and Edinburgh: T. Cadell/W. Davies and W. Creech, 1795), eds W.P.D. Wightman and J.C. Bryce (Indianapolis: Liberty Fund, 1982), pp. 33–105

Adam Smith, *The Theory of Moral Sentiments* (1759), eds D.D. Raphael and A.L. Macfie (Indianapolis: Liberty Fund, 1984)

Tobias Smollett, *The Expedition of Humphry Clinker* (1771), ed. Lewis M. Knapp (Oxford: Oxford University Press, 1984)

The Collected Works of Mary Somerville, 9 volumes, ed. James A. Secord (Bristol: Thoemmes Press, 2004)

Mary Somerville, *Mechanism of the Heavens* (London: John Murray, 1831)

Queen of Science: Personal Recollections of Mary Somerville, ed. Dorothy McMillan (Edinburgh: Canongate, 2001)

Germaine de Staël, *De l'Allemagne* (1813), intro. Simone Balayé (Paris: Garnier-Flammarion, 1968)

Emma Wallington, 'The physical and intellectual capacities of woman equal to those of men', *Anthropologia*, 1 (1874), 552–65, in Barbara T. Gates, ed., *In Nature's Name: An Anthology of Women's Writing and Illustration, 1780–1930* (Chicago and London: University of Chicago Press, 2002), pp. 32–47

William Whewell, 'Review of On the Connexion of the Physical Sciences by Mrs Somerville', *Quarterly Review*, LI (1834), 54–68

The Works of Mary Wollstonecraft, eds Janet Todd and Marilyn Butler, 7 volumes (London: Pickering and Chatto, 1989)

PERIODICALS

European Magazine and London Review
Gentleman's Magazine

Bibliography

The Ladies' Diary
New Annual Register
The Scientific Magazine, and Freemason's Repository

SECONDARY SOURCES

New Oxford History of Music, Volume VI: Concert Music 1630–1750, ed. Gerald Abraham (Oxford: Oxford University Press, 1986)

Margaret Alic, *Hypatia's Heritage: A History of Women in Science from Antiquity to the Late Nineteenth Century* (London: The Women's Press, 1986)

Marina Benjamin, ed., *A Question of Identity: Women, Science, and Literature* (New Brunswick, NJ: Rutgers University Press, 1993)

Marina Benjamin, ed., *Science and Sensibility: Gender and Scientific Enquiry 1780–1945* (Oxford: Basil Blackwell, 1991)

Morris Berman, *Social Change and Scientific Organisation: The Royal Institution, 1799–1844* (London: Heinemann, 1978)

T.C.W. Blanning, *The Culture of Power and the Power of Culture: Old Regime Europe 1660–1789* (Oxford: Oxford University Press, 2002)

John Brewer, *The Pleasures of the Imagination: English Culture in the Eighteenth Century* (London: HarperCollins, 1997)

Frank Brown, *Caroline Herschel as a Musician* (Bath: William Herschel Museum, 2000)

Margaret Bullard, 'My Small Newtonian Sweeper – Where is it Now?', *Notes and Records of the Royal Society of London*, 42.2 (1988), 139–48

Günther Buttmann, *The Shadow of the Telescope: A Biography of John Herschel* (1970), trans. E. J. Pagel (Guildford and London: Lutterworth Press, 1974)

Agnes M. Clerke, *The Herschels and Modern astronomy* (London: Cassell and Company, 1895)

Alice Corkran, *The Romance of Women's Influence* (London: Blackie and Son Limited, 1906)

Shelley Costa, 'The *Ladies' Diary*: Gender, Mathematics, and Civil Society in Early-Eighteenth-Century England', *Osiris*, 17 (2002), 49–74

Uriel Dann, *Hanover and Great Britain 1740–1760: Diplomacy and Survival* (Leicester: Leicester University Press, 1991)

Die Musik in Geschichte und Gegenwart, Sachteil 4: Hamm-Kar (Kassel: Bärenreiter-Verlag, 1996)

Aileen Douglas, 'Popular Science and the Representation of Women: Fontenelle and After', *Eighteenth-Century Life*, 18.2 (1994), 1–14

Cyril Ehrlich, *The Music Profession in Britain Since the Eighteenth Century: A Social History* (Oxford: Clarendon Press, 1985)

Patricia Fara, 'Portraying Caroline Herschel', *Endeavour*, 26.4 (2002), 123–4

Patricia Fara, *Pandora's Breeches: Women, Science and Power in the Enlightenment* (London: Pimlico, 2004)

Paula Findlen, 'Science as a Career in Enlightenment Italy: The Strategies of Laura Bassi', *Isis*, 84 (1993), 441–69

John Gagliardo, *Germany Under the Old Regime 1600–1790* (London and New York: Longman, 1991)

Barbara T. Gates and Ann B. Shteir, eds, *Natural Eloquence: Women Reinscribe Science* (Madison and London: University of Wisconsin Press, 1997)

Jan Golinski, *Science as Public Culture: Chemistry and Enlightenment in Britain, 1760–1820* (Cambridge: Cambridge University Press, 1992)

Dena Goodman, *The Republic of Letters: A Cultural History of the French Enlightenment* (Ithaca and London: Cornell University Press, 1994)

Harriet Guest, *Small Change: Women, Learning, Patriotism, 1750–1810* (Chicago and London: University of Chicago Press, 2000)

Selections from The Female Spectator: Eliza Haywood, ed. Patricia Meyer Spacks (New York and Oxford: Oxford University Press, 1999)

Bridget Hill, *Women, Work and Sexual Politics in Eighteenth-Century England* (Oxford: Oxford University Press, 1989)

Bridget Hill, *Women Alone: Spinsters in England, 1660–1850* (New Haven and London: Yale University Press, 2001)

Marie-Claire Hoock-Demarle, *La femme au temps de Goethe, 1749–1832* (Paris: Stock/Laurence Pernoud, 1987)

Michael Hoskin, *The Herschel Partnership as viewed by Caroline* (Cambridge: Science History Publications Limited, 2003)

Michael Hoskin, 'Vocations in Conflict: William Herschel in Bath, 1766–1782', *History of Science,* 41.3 (2003), 315–33

Michael Hoskin, 'Caroline Herschel: Assistant Astronomer or Astronomical Assistant?', *History of Science,* 40.4 (2002), 425–44

Olwen Hufton, *The Prospect Before Her: A History of Women in Western Europe. Volume I: 1500–1800* (London: Fontana Press, 1997)

Olwen Hufton, *Europe: Privilege and Protest 1730–1789*, 2nd edition (Oxford: Blackwell, 2000)

David W. Hughes, 'Caroline Lucretia Herschel – comet huntress', *Journal of the British Astronomical Association*, 109.2 (1999), 78–85

Rob Iliffe and Frances Willmoth, 'Astronomy and the Domestic Sphere: Margaret Flamsteed and Caroline Herschel as Assistant-Astronomers', in Lynette Hunter and Sarah Hutton, eds, *Women, Science and Medicine 1500–1700: Mothers and Sisters of the Royal Society* (Stroud: Sutton Publishing Limited, 1997), pp. 235–65

Ian Inkster and Jack Morrell, eds, *Metropolis and Province: Science in British Culture, 1780–1850* (London: Hutchinson and Co., 1983)

Colin Jones, *The Great Nation: France from Louis XIV to Napoleon* (London: Penguin, 2003)

Henry C. King, *The History of the Telescope* (London: Charles Griffin and Company, 1955)

Lisbet Koerner, 'Women and Utility in Enlightenment Science', *Configurations*, 3.2 (1995), 233–55

Constance A. Lubbock, *The Herschel Chronicle: The Life Story of William Herschel and his Sister Caroline Herschel* (Cambridge: Cambridge University Press, 1933)

Sir Patrick Moore, *Caroline Herschel: Reflected Glory* (Bath: William Herschel Museum, 1988)

Kathryn A. Neeley, 'Women as Mediatrix: Women as Writers on Science and Technology in The Eighteenth and Nineteenth Centuries', *IEEE Transactions on Professional Communication*, 35.4 (December 1992), 208–16

John North, *The Fontana History of Astronomy and Cosmology* (London: Fontana Press, 1994)

Marilyn Bailey Ogilvie, 'Caroline Herschel's Contributions to Astronomy', *Annals of Science*, 32 (1975), 149–61

Roberta J.M. Olson and Jay M. Pasachoff, *Fire in the Sky: Comets and Meteors, the Decisive Centuries in British Art and Science* (Cambridge: Cambridge University Press, 1998)

Peter Petschauer, *The Education of Women in Eighteenth-Century Germany: New Directions from the German Female Perspective. Bending the Ivy* (Lampeter: Edwin Mellen Press, 1989)

Patricia Phillips, *The Scientific Lady: A Social History of Women's Scientific Interests, 1520–1918* (London: Weidenfeld and Nicolson, 1990)

Roy Porter, *English Society in the Eighteenth Century,* revised edition (Harmondsworth: Penguin, 1991)

Jane Rendall, *The Origins of Modern Feminism: Women in Britain, France and the United States, 1780–1860* (Basingstoke: Macmillan, 1985)

G.S. Rousseau, 'Science books and their readers in the eighteenth century', in *Books and their Readers in Eighteenth-Century England,* ed. Isabel Rivers (Leicester: Leicester University Press, 1982), pp. 197–255

New Grove Dictionary of Music and Musicians, Volume 8, ed. Stanley Sadie (London: Macmillan, 1980)

Sara J. Schechner, *Comets, Popular Culture, and the Birth of Modern Cosmology* (Princeton, NJ: Princeton University Press, 1997)

Londa Schiebinger, *The Mind Has No Sex?: Women in the Origins of Modern Science* (Cambridge, MA and London: Harvard University Press, 1989)

Science Writing By Women, 7 volumes, ed. Bernard Lightman (Bristol: Thoemmes Press, 2004)

James A. Secord, *Victorian Sensation: The Extra-ordinary Publication, Reception, and Secret Authorship of Vestiges of the Natural History of Creation* (Chicago and London: University of Chicago Press, 2000)

Robert B. Shoemaker, *Gender in English Society 1650–1850: The Emergence of Separate Spheres?* (London and New York: Longman, 1998)

Ann B. Shteir, *Cultivating Women, Cultivating Science: Flora's Daughters and Botany in England 1760–1860* (Baltimore and London: Johns Hopkins University Press, 1996)

Margaret E. Tabor, *Pioneer Women*. Fourth Series: Caroline Herschel, Sarah Siddons, Maria Edgeworth, Mary Somerville (London: The Sheldon Press, 1933)

A.J. Turner, *Science and Music in Eighteenth-Century Bath: An Exhibition in the Holburne of Menstrie Museum, Bath, 22 September 1977–29 December 1977* (Bath: University of Bath, 1977)

Amanda Vickery, 'Golden Age to Separate Spheres? A Review of the Categories and Chronology of English Women's History', *Historical Journal,* 36.2 (1993), 383–414

Alice N. Walters, 'Conversation Pieces: Science and Politeness in Eighteenth-Century England', *History of Science*, 35 (1997), 121–54

Neal Zaslaw, ed., *The Classical Era: From the 1740s to the end of the eighteenth century* (Basingstoke: Macmillan, 1989)

Index

accidents
    C.H. 142–3
    W.H. 142
accounts 130
Alexander, William 146
Algarotti, Francesco 94
alien status 81
ambition 7–9, 221
    fulfilled 11–12
    thwarted 93–4, 98–9, 170
annuity 197
apprenticeship 97–8, 142
    'assistant
        Astronomer'133–4
    lifelong 98
arithmetic 31, 32, 85, 230
Astell, Mary 33–4
Astronomical Society of
    London *see* Royal
    Astronomical Society
astronomy 10, 47–8, 135
    C.H. defends 215
    C.H. discontinues 214
    C.H. enjoys 135–6
    C.H. introduced to 51–2,
        119–20, 130–1
    C.H. place in 167
    C.H. retains contact with
        205–6

developing interest of
    W.H. 72–3, 92–3, 109
and embroidery 48
importance to Herschels
    107–8
pursued by women 94
Atlas Coelestis 184
Aubert, Alexander 1, 117,
    148, 151
Austen, Jane 29–30
awards and recognition
    209–10
    honours and honorary
        memberships 212,
        213–14

Baillie, Joanna 221
Baily, Francis 187
Banks, Joseph (Sir) 117, 145,
    150, 221
Bassi, Laura 193
Bath 59, 61, 72
    C.H. invited to 62–3
    C.H. travels to 71–2
    and consumerism 75–6
    and millinery 113
    and musicians 76, 77–8
    offers little to C.H. 87
    ostentation 74–5

Bath (*continued*)
  population 77
  and slavery 75–6
  as spa resort 73–4
  W.H. settles in 59, 61
'Beau Nash' 76
Beckedorff (Mrs) 56–7, 191
Behn, Aphra 94, 95
Birmingham 104, 105, 106
birth 14
Blagden, Charles (Dr) 1,
  149, 150
'Book of Work Done' 1
'Books of Observations' 137
botany 34
Bristol 109, 110–11
British Catalogue 184
  amended by C.H. 186, 189
  unreliability 185
*British Public Characters of
  1798* 166–7, 192
Bulman family 81, 89
Bulman (Mrs) 84
Burney, Charles 122–3
Burney, Frances 90–1, 153,
  171, 191–2
  describes C.H. 179

Cantelo (Miss) 103, 107–8
Cape of Good Hope 206, 248
Carter, Elizabeth 94, 95
catalogue of nebulae (C.H.)
  147, 201, 202–3

catalogue of nebulae and
  clusters (C.M.) 136
*Catalogue of Stars, taken from
  Mr Flamsteed's
    Observations ...* 186–8
  assessed 188
  C.H. prepares revised list
    187–8
  C.H.'s achievement 188–9
  objections to 187–8
catalogues 119
Cavallo, Tiberius 150
celestial holes 206, 260
character *see* personality
Charlotte (Queen) 159
Châtelet, Emilie du 193
chemistry 33
childhood 13–59
  confirmation 39, 40–1
  loveless 20–1
  unsociability 227
Christian Religion (*The*)
  33–4
Clay Hall 145
Clerke, Agnes M. 5–6
clothing 28–9
Colnbrook (Mrs) 90
comets
  C.H. sweeps for 133–4
  discovered by C.H. 163–4,
    181, 209
  discussed in press 165–6
  finding announced 164

first 1, 147–8
last 189
second 163–4
early discoveries 150
early theories on 151–2
'Commonplace Book' 135
communication between
    C.H. and W.H. 173
Copley Medal 117
Crab nebula (M1) 136

D'Arblay (Madame) *see*
    Burney, Frances
Datchet 125–6, 200 cost of
    living 128–9
Day-Book (The) 190, 191,
    194
De l'Allemagne 44
death (of C.H.) 11, 216
depression 194, 205
*Description of a Forty Feet
    Reflecting Telescope* 183–4
Deutsche Chronik 23
disfigurement 43
domestic servants
    hierarchy 84
    lack of 234
    salaries 157
Dreyer, J.L.E. 188–9
duty 45

earnings *see* salary
eclipse, solar 50–1

Edgeworth, Maria 33, 154
education 7
    in Britain 27–8
    of C.H.
        deficencies in 62
        self-training 8–9, 92
        'useful' elements allowed
            53
    curriculum 31–2
    in Germany 27, 30, 31, 230
    as means of achievement
        55
    of women 29–30, 31,
        32–3, 37, 222–3
    mathematics 33,
        35–6,130
    science 47–8
    see also schooling
Emile 36
Encke, Johann Franz 209
Englefield, Harry (Sir) 165
English language 82–3, 85
English people 99
entomology 34
epigraph (on C.H.) 216–17
European Magazine 152
Evelina 90–1
*Expedition of Humphry
    Clinker (The)* 74–5

Fara, Patricia 6–7
'female astronomer' 159, 160,
    165–6

Female Rights vindicated or the Equality of the Sexes Morally and Physically proved 47, 48

Female Spectator (The) 34, 36

female suffrage 222

feminist icon 222–3

First Autobiography (The) 20–1, 58, 91, 97, 99, 101, 102, 104, 112, 142, 160, 163, 195

Flamsteed, John 185

Fleming, Anne 101

Fontenelle, Bernard le Bovier de 94, 95

French language 52, 59

friendships 65–6, 88–9
    lack of 19

future prospects (C.H.'s) 61–2

Gentleman's Magazine 42, 197

George I 24, 26–7

George II 17, 24

George III
    grants W. H. money 144–5
    inspects telescope 175
    pays C.H.'s salary 155–6
    summons W.H. to court 123
    views on Germany 25

Georgium Sidus ('George's Star') 244

Germans 23–4

Germany 22, 24
    divorce in 44–5
    education 27, 30, 31, 230
    as viewed by Europeans 22–3

glass 115–16

Gold Medal (Prussia) 2, 212

Gold Medal (Royal Astronomical Society) 2, 207

governess 156–7

Handel's music 100, 102

Hanover 14, 16
    C.H. leaves 70–1
    C.H. returns to 21–2, 196, 197–8
        discontinues astronomy 214
    contempt for population 21–2
    as headless state 24, 25–6, 64–5, 229
    lack of development 26–7
    occupied by French troops 25

harpsichord 86

Harrison (Mrs) 103, 107–8

Herschel, Abraham 15

Herschel, Alexander 62, 83, 87–8, 114, 132
    helps W.H. 118–19, 120, 121

lives with W.H. 81
musical career 61, 81
returns to Bath 125, 126, 130
Herschel, Anna 14, 15
  aids C.H.'s millinery studies 56–7
  blocks C.H.'s ambitions 52, 53, 54
  compensated by W.H. 70–1
  distrusts learning 54–5
  employs servant 71
  exploits C.H. 17–19, 20, 38
  marries I.H. 16
Herschel, Arabella 3–4
Herschel, Caroline
  birth 14
  death 11, 216
Herschel, Dietrich 20, 44, 58, 61, 195–6, 197
  belittles astronomy 215
  musical abilities 65
  receives money from C.H. 196
Herschel family
  alternative names 226
  appeased by C.H. 69–70
  C.H. sacrificed to 59
  relationship between siblings 170–1
  communication 172–3
Herschel, Frantz Johann 42

Herschel, Isaac 14, 15, 16
  and children's abilities 17, 50
  and children's future 46
  introduces Caroline to astronomy 51–2
  marries Anna Ilse Moritzen 16, 226
  sent to Britain 18
  takes in paying pupils 49
  unsociability 227
  views on marriage 41–2
Herschel, Jacob 14, 15, 61
  attitude to C.H. 50, 67
  attitude to trade 56
  blocks C.H.'s ambitions 52, 66–7
  C.H. attends to 50
  lifestyle 45–6, 49–50
  musical abilities 46, 65
  sent to Britain 18
Herschel, John 12, 134, 194, 199, 248
  C.H. comments on his work 203–5
  and C.H.'s assistance 207
  maps southern hemisphere 205, 206, 248
  receives Gold Medal 207
  recognises C.H.'s strength 202
Herschel, Mary *see* Pitt, Mary

Herschel, Sophia 14, 16, 19, 20, 226–7

Herschel, William
  assisted by C.H. 98–9, 119, 132–3, 207
    as amanuensis 140–1
    C.H.'s ceaseless support for 5, 14–15, 105
    C.H.'s contribution vital 144
    grinds mirrors 120–2
    selfless devotion 93–4
    to her own disadvantage 148–9
  boasted of progress 118
  death 195
  discovers Uranus 2, 110, 115
  employment for brothers 61
  encourages C.H. 131, 135–6
    gives credit 183–4, 185
    shows faith in 66
    teaches singing 85–6
  health 194
  income 123
  marriage 160–1
  rivalry with Linley 242
  sent to Britain 18
  teaches music 122–3
  travels to Göttingen 146
  welcomes Caroline home 18

Herschels and Modern Astronomy (The) 5–6

Historia coelestis Britannica 185

history of science 9–11
'hobbies' 48–9
Hof- und Töchterschule 31
house moving 93, 109–10, 145, 170, 189–90
household duties 13–14, 58–9, 83–4, 84–5, 88, 124
  C.H. taught by Mrs Bulman 84
  housework 17, 20, 38, 40
  importance diminishing 167
  on low income 128
  prevent marriage 43–4
  responsibility of women 129–30
  shopping 82–3, 128, 129
  training in 98

Humboldt, Alexander von 212

humour, sense of 198

independence 8, 9, 55, 217, 219
  compared with other women 156–7
  salary 155–60
  thwarted by W.H. 99

Judas Maccabeus 102

Karsten (Mademoiselle) 55–6
King of Prussia 2, 212
knitting 38, 49, 69
Kratzenstein, Christian
    Gottlieb 150

La Roche, Sophie von 22–3,
    153–5
    describes C.H. 178–9
Ladies' Astronomy 192–3
Lalande, Joseph Jérôme de 2,
    165, 174, 192–3
Lamballe, Marie (Princesse
    de) 174, 253–4
Laplace, Pierre-Simon 96
'Learned Lady's' 198–9
learning *see* education
*Life and Adventures of Miss
    Caroline Herschel Solely for
    the Amusement of Lady
    Herschel* 3
Linley, Thomas 104, 111, 242
literacy 38–9
    levels in Europe 27
London 72–3, 90–2, 117
Lysons, Samuel 110

Mahon, Elizabeth 102–3
Malebranche, Nicolas 35–6,
    37
Marchioness of Lothian 106,
    123
marriage 39–40, 44

marriage (of C.H.) 41–2
    lack of prospects 43–4, 232
marriage (of W.H.) 160–1
    affects C.H. 161
        causes frustration 169–70
        problems 167–8
Marsh, John 132
Martin, Benjamin 95
Maskelyne, Nevil 110, 117, 118
    and C.H.'s *Catalogue of
        Stars …* 186, 187–8
    on C.H.'s discovery 151
    on C.H.'s telescope 138–40
mathematicians 182
mathematics 10, 15, 31, 134
    and astronomy studies
        47–8
    studied by C.H. 31, 135,
        230
    in women's education 33,
        35–6, 130
Méchanique Céleste 96
'mediatrix' (scientific) 181
memoirs 3–4
    fictional treatment 4–5
Messiah 100
Messier, Charles 136, 137,
    150–1, 164
millinery 56–7, 112–13
    collapse of business 113–14
mirrors
    C.H. works on 118–19,
        120–2, 140, 141

mirrors (*continued*)
    difficult construction
        176–7
    use in telescopes 115, 116
    W.H. works on 120–1
modesty 47
Moritzen, Anna Ilse *see*
    Herschel, Anna
music 15, 16
    in C.H.'s life
        copies scores 100
        directs trebles 112
        first stage appearance
            100, 101
        last performance 122
        relinquished 125–6
        singing
            career thwarted 98–9
            oratorios and concert
                songs 109
            principal soloist
                103–4, 111
            voice training 67–9,
                85–6, 92
        standard 241–2
        'tryal' of abilities 62,
            63–4, 66, 70
        turns down invitation
            104–6
        values own abilities 107
        violin playing 49, 65, 66
    as profession or trade 64,
        79

in W.H.'s life
    abilities 61, 65
    benefit concert 103–4,
        241–2
    compositions 123
    established 61
    last performance 122
    organist 79–80
    precarious career 78–9
    reputation 111
musicians 77–8
    employment 78
    lifestyle 64–5, 127
    personality 127

Nash, Richard ('Beau Nash')
    76
natural philosophy *see* science
nebulae 1, 200–1, 209
    catalogues 136, 201
    discovered by C.H. 137–8
    distinguished from comets
        137
    number known 137
needlework 20
Newtonianism for the Ladies 94

obituary
    on C.H. 216
    on W.H. 197
Octagon Chapel 79–80
    choir 100
    W.H. organist 79–80

Olbers, Heinrich 215
old age 3, 11, 199–217
  and astronomy 199–200
  discontinued 214
  mental alertness 215
On the Connexion of the
  Physical Sciences 210
optical industry 116, 244

Pandora's Breeches 6–7
Papendiek (Mrs) 129, 146,
  167, 168
  describes C.H. 178
personality (of C.H.) 5, 6,
  89–90, 126–7, 198
  as ideal scientist 179,
  182–3
Philosophical Transactions
  183
physics 47, 94–5
Piery (Madame du) 193
Pigott, Edward 186–7
Pitt, Mary 160–1, 167–8
  and W.H.'s affections
  168–9
Playfair, John 174
Plurality of Worlds 94, 95
poets 181–2
press (popular) 165–6
public events 198–9
public (general) 152
published work 183
Purinina (Countess of) 193

Recherche de la vérité 35–6
Reduction and Arrangement
  in the Form of a Catalogue,
  in Zones, of all the Star-
  clusters and Nebulae
  observed by Sir W. Herschel
  in his Sweeps (*The*) 207
reflector 144–5
religion 96, 97
reputation recognised 217,
  219
Rousseau, Jean-Jacques 36
royal astronomer 123
Royal Astronomical Society
  2, 11, 206–7
  honorary membership
  209–10, 211–12
Royal Institution 220
Royal Irish Academy 2, 212,
  213–14
Royal Society 117, 186, 212
royalty 173–4, 192

salary paid by George III 1,
  155–6
  requested by C.H. 157–8
  requested by W.H. 158–60
Saturn 115
schooling 37–8, 46–7
  graduation 39–40, 41
  see also education
Schubart, Christian Daniel
  23

science 123–4
    amateurs and profes-
        sionals 124, 220
    elevates mind 96–7
    financial reward 124, 157
    individual disciplines 180,
        261
    as masculine subject 180
    popularisation 95, 220
    and religion 96, 97
    as sociable activity 94–5
    in women's education 33,
        34, 35–7, 47–8
scientists 179, 181, 182–3
Second Autobiography (The)
    18
self-improvement 46–7, 55,
    86
    distractions from 49–50
self-sufficiency 45
Seven Years War 17–18, 21, 25
sewing 48, 55
sexual maturity 39–40, 41
sight 191
Slough 145, 168
    High Street lodgings 190
smallpox 42, 43
Smith, Adam 150, 181–2
Smollett, Tobias 74–5, 77
social calls and gatherings
    43–4, 191
social graces 90–1, 92, 101, 112
    W.H.'s list 89–90

Solar eclipse 50–1
Somerville, Mary 28–9,
    96–7, 209–12
South, James 207–9, 213
    on C.H. as astronomer
        208–9
    on C.H.'s assistance to
        W.H. 207–8
spinning 48
Staël, Germaine de 44–5
stars
    arranged in zones 201
    found by C.H. 186
St-Fond, Barthélemi Faujas
    de 171–3
stillness (stillsitzen) 47
subservience 7
Swift, Jonathan 24

tables 119
talismans, collecting 154
telescopes 136–7
telescopes (C.H.'s)
    access restricted 190
    improved 138, 139–40
    time away from 141
    and visitors 146
telescopes (W.H.'s) 140
    40 feet long 174–5
        concert held in 175
        construction difficul-
            ties 176
        mirror 176–7

popular attraction
174–5, 178
turning of tube 177
unsuccessful 177–8
weight supported 177
construction
accidents 142–3
C.H. assists 92–3, 97
largest 144–5
reflectors 115, 117–18
size and power 114–15
and visitors 110, 116–17,
173–4
theatre 57
Theory of Moral Sentiments
181–2
*third catalogue of the
comparative brightness of
the stars (A)* 184–5

Upton 167, 168
Uranus 2, 110, 115
usefulness 45, 63, 185, 199,
223

*Vindication of the Rights of
Men (A)* 44
*Vindication of the Rights of
Woman (A)* 7–8, 27–8
visitors 110, 146, 155
disturb C.H. 116–17, 200
look through telescopes
116–17, 173–4

note C.H.'s part in
discoveries 173
pay homage to C.H. 148,
149
royalty and aristocracy
173–4
voice training 85–6
'gag' method 68, 86
helped by knowledge of
violin 68–9
and household tasks 68
improvement in range 92

Walcot 93
Wallington, Emma 222–3
Watson, William (Sir) 121,
124–5
weaving 48
Whewell, William 210, 261
window tax 116
Wollstonecraft, Mary 7–8,
157
on education 27–8, 33
on marriage 44
women
achievements 152–3
education 29–30, 31,
32–3, 37, 222–3
for domesticity 32–3
mathematics 33, 35–6,
130
science 35–7, 47–8
tertiary 221

women (*continued*)
  opportunities lacking 216
  scientists 180, 192–3
    achievements 221
    ambition 221
    participation 219–20
    'spirit of inquiry' 193

work, importance of 200, 202

Young Gentlemen's and Ladies Philosophy (*The*) 95

ICONSCIENCE

THE ICON SCIENCE 25TH
ANNIVERSARY SERIES IS A
COLLECTION OF BOOKS ON
GROUNDBREAKING MOMENTS
IN SCIENCE HISTORY, PUBLISHED
THROUGHOUT 2017

The Comet Sweeper
9781785781667

Eureka!
9781785781919

Written in Stone
9781785782015
(not available in North America)

Science and Islam
9781785782022

Atom
9781785782053

An Entertainment for Angels
9781785782077
(not available in North America)

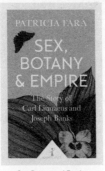

Sex, Botany and Empire
9781785782275
(not available in North America)

Knowledge is Power
9781785782367

Turing and the Universal Machine
9781785782381

**Frank Whittle and the
Invention of the Jet**
9781785782411

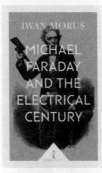

**Michael Faraday and the
Electrical Century**
9781785782671

Moving Heaven and Earth
9781785782695